AF449130

Business and Society
General editor   Alex J. Robertson

Ferranti and the British electrical industry, 1864-1930

J. F. WILSON

Manchester University Press

Manchester and New York
Distributed exclusively in the USA and Canada
by **St. Martin's Press**, New York

Copyright © J.F. Wilson 1988

Published by Manchester University Press
Oxford Road, Manchester M13 9PL, UK

Distributed exclusively in the USA and Canada by
St. Martin's Press, Inc., 175 Fifth Avenue, New York, NY 10010, USA

British Library cataloguing in publication data
Wilson, J.F.
   Ferranti and the British electrical
   industry, 1864–1930. — (Business and
   Society).
   1. Ferranti, Sebastian Ziani de
   2. Electric engineers — Great Britain
   — Biography
   I. Title   II. Series
   338.4'762131'0924   TK140D4

Library of Congress cataloging in publication data applied for

ISBN 0 7190 2368 8 *hardback*

With thanks to Ted Musson

Typeset in Great Britain by
Williams Graphics, Abergele, North Wales

Printed and bound in Great Britain by
Biddles Ltd, Guildford and King's Lynn

Contents

General introduction

Business history is an old-established branch of history by now, having emerged in Britain during the early years of this century. Manchester, in which this present series of books was planned and is being published, had an important role in the process of emergence, through the work of men like George Unwin and Arthur Redford, successively Professors of Economic History in Manchester University from 1910 to 1960. But for all its long and respectable pedigree, business history has not enjoyed the prominence it arguably deserves among the many facets of historical studies. It has remained largely a field for academic specialists rather than for students or readers outside institutions of higher learning. The subject has generated a substantial body of literature over the years, but it has only been very occasionally that a work of business history has made an impact outside the ranks of the professional initiates. And yet, business history ought to have a wider audience, both within and beyond the boundaries of formal higher education. For we have all been, and continue to be, profoundly affected in our tastes, comfort, economic security and social habits by business corporations, individual businessmen, and the decisions and choices they make. Business-men, indeed, have exercised in their way an influence on modern economic and social life no less significant than that of the statesmen and soldiers, politicians and trade-union leaders whose lives and activities are conventionally recognised as the proper stuff of history, particularly at the popular level.

It is the intention of the *Business and Society* series to bring business history to a wider readership than it has usually reached. We trust, of course, that the books in the series will be of interest and use to professional business historians, especially for the purpose of undergraduate teaching. But it is our particular hope that the books will appeal to readers who, without being business history specialists, have an interest − either professional or purely personal − in the history of modern society, its problems and its achievements. As the title of the series indicates, these are not intended to be studies of business activity narrowly defined. Their focus is not directed inwards, at the internal workings of business enterprise. Such matters clearly have their place in the studies, but the books will devote much of their attention to the broader context and significance of business activity, as a powerful determinant of the general course of economic development, of patterns of social life, of public taste, of state and local government policy, and so on. The intention, in short, is to illuminate the role of businessmen and business institutions in forming the modern social fabric.

In pursuing this aim, some of the studies will (like this one) adopt the biographical approach. Thomas Carlyle's dictum, 'The history of the world is but the biography of great men', applies as surely to the business branch of history as to any other. Subsequent books will turn the spotlight on to groups of businessmen with common interests; the pioneers of aircraft manufacture in Britain, for example, or the ship-builders. Yet others will examine corporations rther than individuals, and analyse their relationships with the societies within which they functioned. But whatever approach any individual volume adopts, the books in the *Business and Society* series are intended to combine sound scholarship with straightforward, readable presentation.

John Wilson's study of Sebastian Ziani de Ferranti fulfils that intention amply. The book is based on several years of research in the Ferranti company and family archives. It clearly demonstrates its subject's significance as a pioneer in the development of a public utility which we all now take for granted, but without which the economic and social life of the country would be very different: electricity supply. Indeed, Dr Wilson makes it plain that Ferranti's importance as a pathfinder in the development and exploitation of electricity was manifold. He played key roles in the development of the technology of electricity generating and in the foundation of the modern system of public electricity supply, culminating towards the end of his life in his contribution to the establishment of the National Grid which remains at the heart of Britain's electricity supply arrangements. Indeed, it might well be argued that Ferranti alone among British electrical engineers and manufacturers was a force to be reckoned with at an international level before 1914, in an industry dominated by enormous German and American concerns. But Ferranti's activities did not stop at the supply side of the electrical industry: for after the First World War, his firm took a leading part in the growth of the electrical and electronic appliance industry serving the consumer market.

British businessmen of Ferranti's generation have been fiercely castigated for their ignorance of matters scientific (a result, it is supposed, of their classical public-school educations), for their failure to exploit products that embodied new technologies, for their unwillingness to invest in large-scale plants embracing the most efficient production methods, and for various other failings. The shortcomings of the late Victorian and Edwardian entrepreneur, indeed, are widely seen as a major factor in starting Britain's decline as a power in the world economy. A considerable literature has grown up around this matter, and Dr Wilson's book may be seen as contributing significantly to the debate. It is obvious that Ferranti does not readily conform to the usual sorry stereotype. *His* public school, for instance, actively encouraged his scientific bent; and when he had exhausted the school's facilities, it found him extra-mural help. And far from trailing behind the most advanced technology of the day in his chosen field, Ferranti persistently operated at and even (Dr Wilson contends) beyond the leading edge of the contemporary practical technology.

Yet Ferranti had his blind spots too. Against his brilliance as an inventor and innovator, for example, there clearly has to be set his conservatism in matters affecting the financial and administrative control of the company that bore his name. This constantly weakened its ability to stand in competition with such giants as the German AEG or the American GEC, and finally brought the firm to the brink of dissolution in 1903. In this, perhaps, Ferranti was more typical of his generation than he was in other respects. But whatever his failings in matters of finance and company administration, readers of Dr Wilson's account cannot fail to be convinced of Sebastian de Ferranti's importance as a force in shaping modern Britain's industrial and domestic life.

Alex J. Robertson
(General Editor)
Manchester
November 1987

Preface and acknowledgements

Living in a period when electricity is such an intrinsic feature of our world, it is difficult to envisage life without it. Today, in Britain, power stations and pylons dominate the landscape, and electricity is transmitted at 400,000 V over the 'supergrid' to approximately twenty-one million customers. One hundred years ago the situation looked quite different, and the creation of an electricity supply industry capable of ousting its major competitors from the markets for lighting (gas) and industrial power (steam) took several decades. A large number of entrepreneurs contributed to this outcome, one of the most interesting of whom was Sebastian Ziani de Ferranti. Spanning the period between the emergence of the British electrical industry in 1882 and the implementation of the momentous 1926 Act creating a national electricity network, his career illustrates the technological, legal and commercial problems this new industry had to overcome in its struggle for recognition in Britain as a vital component of a modern economy. Ferranti was deeply involved both in the development of an efficient supply industry and in the establishment of a viable electrical engineering industry, advocating in word and deed what became the two main strands of his philosophy: the potential of having cheap electricity for domestic and industrial consumers in an 'all electric' age, and the need to improve the technological awareness of British industry. This review of Ferranti's life will provide clear illustrations of these motivating forces, from his construction of the world's first modern power station at Deptford between 1888 and 1891 to the emergence of Ferranti Ltd as one of the most progressive British electrical firms. He was regarded as a leading pioneer by contemporaries, and several honours were bestowed upon him by the scientific fraternity in recognition of a lifetime committed to the industry. These honours included being invited to remain as President of the Institution of Electrical Engineers for two consecutive terms, the first person ever to be in such a coveted position, and in 1927 being made a Fellow of the Royal Society. One could never argue that he was solely responsible for the establishment of the electrical industry, and it will become clear that there were many who played equally important roles, but his contributions in the supply and manufacturing fields were to prove of lasting significance in the twentieth century.

The years 1880–1930 mark out a vital period in British economic history. It was a time when the ability of businessmen to compete with foreign industrialists was brought sharply into focus by the emergence of new technologies. Ferranti was a prolific inventor, taking out a total of 220 patents in forty-nine years on such varied subjects as alternators, meters, cables, steam engines, stop valves, textile machinery, gyroscopes and radio components. His attempts to canvass support for these ideas illustrate the problems of pioneering on the frontiers of technology in this crucial period. The fortunes of the succession of businesses in which he became involved also reflect accurately the vicissitudes experienced by the British electrical engineering industry. However, by insisting on a far-sighted policy of investing a high proportion of turnover in ambitious development projects, Ferranti ensured that the product range remained competitive. He was particularly concerned that foreign competition was harming such technologically progressive industries, and he argued vigorously for a more scientific approach towards design and product development to improve his company's chances of survival. Another means of ensuring the profitability of British electrical manufacturers was to improve the degree of cohesion within the industry, and he worked actively to establish a powerful trade association capable of controlling prices and reducing the more harmful effects of competition. Not that he ever encouraged concentration within the industry; he was a firm believer in the individualist philosophies which were such a pervasive influence on businessmen up until the Second World War. Ferranti was very much a man of his time in that respect, but this did not prevent the pursuit of technological excellence, and he encouraged British industry in general to be more ambitious in its adoption of new ideas. Indeed, the company he established has remained a vital component in Britain's technological vanguard on the basis of such an approach.

Clearly, a review of Ferranti's life is a review of the struggle to nurture the growth of the electrical industry, and in this book I have attempted to merge the biography with a study of business and industrial history. The first two chapters are concerned with his dynamic first decade as an entrepreneur, during which he established the basic principles of the system which eventually became standard practice throughout the world's electricity supply industries. This is directly linked with the study in Chapter 5 of the State's involvement in the industry, and the establishment of the national grid network, while the intervening chapters analyse the environment for entrepreneurship in late-Victorian and Edwardian Britain. (For ease of reference, a statistical appendix is included to give a clear picture of the company's performance over time.) The aim is to present some idea of Ferranti's contribution to the industry in the forty-nine years of his business career, while describing the company's evolution in general detail. A wide variety of sources have been used, and a bibliography is included, but as no footnotes are carried, only the author's name and the date of publication have been used in the text to indicate the use of any particular book or article listed there.

During the course of this project, I have been given substantial assistance from a large number of people too numerous to mention, but I should particularly like to thank my supervisor and colleague, Professor A. E. Musson, for all his endeavours on my part since it all began in 1978. This book is dedicated to him in recognition of his much-valued advice and guidance. The editor of the series, Mr A. J. Robertson, has also generously given up his time to provide useful hints on the presentation of my research. Other academics who have been of great help include Professor Leslie Hannah and Dr A. J. Marrison, but all blame for any errors naturally rests only with me. I must also record my debt to S. Z. de Ferranti's grandsons, Sebastian and Basil, for giving me access to their archives, while Mr A. K. H. Thomas, Ferranti plc's Publicity Manager, has supportively supervised the project from the company's side throughout. Business historians everywhere will appreciate the value of such work. Successive Ferranti archivists, C. J. Somers, R. J. Campbell (now sadly deceased) and Cliff Wimpenny, have carefully guided me through the available material, ably assisted by Mrs E. Walsh and Mrs D. Scott. Finally, my wife Barbara deserves much credit for the way in which she has endured the whole thing for over eight years, and all my thanks go to her for this forbearance.

JFW

Portrait of Sebastian Ziani de Ferranti, D.Sc., F.R.S., by
D. Jagger, 1931 (*Courtesy of the Institution of Electrical Engineers*)

1

The young inventor – entrepreneur

Reviewing the early life of a man destined to make a series of major contributions to the development of the British electrical industry can be an invigorating experience, but it is all too easy to lapse into a story featuring only those events which affected the contemporary technology directly. One must not be misled by any false sense of inevitability, because, as this chapter clearly illustrates, the young Ferranti's experiences were not altogether encouraging. He received the fullest support from his parents, not to mention an education of great value to a budding engineer, providing an environment conducive to the development of an original mind. Financial support for his schemes was also forthcoming from established businessmen with proven records: convincing evidence of the persuasive nature of his arguments. Yet it was only through committed endeavour that Ferranti remained as enthusiastic as ever about his chosen career, doggedly refusing to bow before any of the legal, commercial or technical problems placed in his path. He remained an optimist, committing his considerable energy and talent to what became a lifelong ideal, the creation of the 'All Electric' age. It is a story which will provide an example of perseverance at a time when, arguably, the general environment failed to provide sufficient encouragement to those with a committed attitude towards business and the advancement of new technologies. Historians have been highly critical of late Victorian and Edwardian businessmen, alleging that, on the whole, there was a failure to innovate and compete effectively against the emerging industrial countries like Germany and the USA. The poor performance of the electrical industry up to 1914 has been used to support this thesis, but it will become clear that a combination of circumstances prevented men like Ferranti from extending the use of electricity in this country, leading to the domination by German and American companies of the British market by the 1890s. The small band of British electrical pioneers were severely handicapped by the lack of market and financial support, and only perseverance and enterprise prevented their complete elimination in the late nineteenth century. Ferranti's career in the 1880s illustrates this story graphically, because without these qualities he could easily have been dissuaded from entering such a difficult industry. By continuing in his chosen career as an electrical engineer, he was to emerge within a decade as one of the most significant British contributors to the technology, laying the foundations for a career built around his desire to see electricity adopted in all walks of life.

It is always fascinating to speculate on the sources of genius, and an examination of Sebastian Ziani de Ferranti's ancestry reveals a succession of colourful, talented and highly original characters. Part, at least, of his character and motivation must have come from stories of his Italian forebears which can be traced back as far as ninth-century Venice. The Ziani name is actually found among the twenty-four houses forming the body of the Patrician nobility, while in 1173 Sebastian Ziani was elected Doge of Venice, and his son, Peter, attained the same position in 1205. These two men reigned during the period when Venice was beginning to reach the peak of its commercial and political power; an indication of the family's standing in Italian politics at this time. It also established the Ziani tradition of states-manship which was continued into the eighteenth century by the Bolognese branch of the family after its move from Venice around 1300. It is from the Bolognese strain that the Ziani de Ferranti family originates and, although politics remained the principal interest of the family, by the nineteenth century certain members were beginning to develop a greater interest in the arts. The first to inherit the Ziani de Ferranti name — the family name had been changed as a result of an inheritance — was Marco Aurelio (born 1801), the grandfather of Sebastian Ziani de Ferranti, who by the age of sixteen was already a highly-regarded virtuoso violinist. He was later to develop an interest in the guitar, at a time when musicians like Weber and Schubert were recognising the potential of the instrument. By the mid-1820s, Marco Aurelio was touring the main European cities exhibiting his rare skills as a guitarist, eventually settling in Brussels, where he took up the post of Professor of Italian Language, Literature and the Guitar at the Royal Conservatoire. Such was his reputation that in the mid-nineteenth century he was regarded as one of the finest guitarists in the world, an achievement rewarded with the honour of being made Guitarist to the King of the Belgians. Clearly, the innovative talents of the Ferranti family were not unique to the more recent generations, and his forebears were always a source of inspiration to our subject.

Although not inheriting his father's musical ability, Marco Aurelio's eldest son, Cesar (born in 1831), was to enjoy a successful career as a photographer. From the mid-1850s he built up a flourishing business in Brussels at at time when this art form was only just emerging from the laboratory. The challenge of developing the photograph as a viable alternative to the painting was pushing this new breed of artist into wide experimentation, and Cesar made his contribution to the technology in 1872 when he patented an improved process of colouring photographic plates. It was while living in Brussels that Cesar met a talented concert pianist, Juliana Szczepanowski, at one of her performances in the Ostend Casino. Juliana was the daughter of a Liverpool portrait painter, William Scott (1797–1862), and in her younger days had been a headstrong character with an aversion to paternal authority. She longed for the excitement and freedom of a musician's life, and after marrying in 1845 a Polish exile, S. Szczepanowski, who aspired to great heights as a guitarist, she embarked on an extensive tour of the major European concert venues. An indication of their lifestyle is that their four children were all born in different cities: Emilka (1846) in Edinburgh, Wladziu (1848) in Dresden, Vanda (1850) in London, and Vincent (1852) in Leipzig. They lived an exciting life, but the call of the Polish cause proved far too powerful for Szczepanowski, and in 1852 he deserted his young family in Ostend, leaving Juliana to support the children through her musical endeavours. This naturally affected her badly, but after meeting Cesar in 1857 she was beginning to regain some of her lost confidence. Cesar was at that time looking for a more lucrative market in which to exploit his talents and, after meeting William Scott in 1860 he decided to move to Liverpool and establish a partnership with the artist, offering both artforms to the wealthy classes of the industrial north-west. It was also over seven years since Juliana had seen or heard from her husband, leaving her legally free to marry Cesar in 1860.

When William Scott died, in 1862, he was content in the knowledge that Juliana was a more settled person and married to a highly successful photographer. Their home and studio in Bold Street, Liverpool, became a thriving concern, and over the next twenty years Cesar was able to generate a good income, particularly after receiving a royal commission to photograph the Princes Leopold and Arthur in 1867. This success was reflected in the purchase of a second house, in Egremont, across the Mersey, where the growing family could live free from the bustle of Liverpool. Although in 1866 Juliana's two sons had joined the merchant navy, in 1864 there had been a further addition to the family when, on 9 April, Juliana gave birth to a son. A month later he was baptised into the Roman Catholic Church and christened Sebastian Pietro Innocenzo Adhemar Ziani de Ferranti. Apart from being a rather weak child, constantly prone to coughs and colds, the young Ferranti was to enjoy a pleasant ten years in Egremont. He was, of course, very much the baby of the family, and he was spoilt by his doting mother and by a nanny hired to relieve Juliana of the domestic drudgery. In fact a very close bond between mother and son was forged in this period, a bond which provided Ferranti with a secure family base on which to build his life.

During the years at Egremont the young Ferranti was often to be seen on the beaches at the head of the Mersey estuary watching the Liverpool mercantile traffic. Living so close to this activity, it is not difficult to understand why ships captivated the young boy's imagination and, in the sketch-books eagerly provided by Juliana, he drew all kinds of vessels. His parents recognised very early that he possessed a natural artistic flair, no doubt anticipating that William Scott's talents might have been passed on, but Ferranti was soon to use this ability in a completely different manner from his grandfather. The book written by de Ferranti and Ince, *The Life and Letters of S. Z. de Ferranti* (1932), recounts this story in great detail, illustrating how the young boy's imagination was fired by technical matters from an early age. Ferranti's fascination with engineering originated with the textbooks acquired by his step-brother, Vincent, while revising for merchant navy officer's examinations. It was at this stage that Ferranti's interest in the size and shape of ships was converted into an instinctive inquisitiveness regarding the reasons why and how prime movers worked. He later recalled, in his address to the Institution of Electrical Engineers (IEE) commemorative proceedings in 1922, that Vincent's influence was absolutely vital in prompting that initial spur to his technical interests, and by the time he was seven Ferranti was already drawing steam engines and locomotives.

The young Ferranti's fascination with steam engines was not confined to drawing or reading, because by the time he had started attending his first school, a preparatory boarding school in Hampstead, in March 1874, he had built up a collection of models. He was particularly impressed with a kit which had to be assembled by the purchaser, because it gave him detailed information about the working components of the engine. His parents also sent such interesting books as Samual Smiles's *Life of the Stephensons* and R. Routledge's *Discoveries and Inventions of the Nineteenth Century*, expanding his knowledge of the mechanical arts. In fact, by the time he had moved up to a senior school, St Augustine's at Ramsgate, it was already apparent that his concern with things mechanical was coming to dominate his life, often at the expense of his academic work. Cesar and Juliana frequently reminded their son that it was essential not to neglect his studies, but his lack of academic application was to frustrate them continually. One need only read de Ferranti and Ince's account (1932) of his early life to gain the impression that his writing was more phonetic than correct. Ferranti, of course, is not the only inventor–engineer who has incurred the wrath of his superiors over this matter, because, as Clark (1977) notes, T.A. Edison, one of the most distinguished men of the age, was called an 'addled youth' by a dissatisfied teacher. Nevertheless, although Ferranti's academic record never proved promising, it is important to remember that it was to be in the practical field that he was to make his greatest contributions.

Despite his academic shortcomings, Ferranti was still a popular pupil at St Augustine's. The teachers recognised a mild-mannered and pleasant disposition in the youth and, in recognition of this responsible approach, they allowed him access to his own workroom in which to pursue certain practical–experimental interests. St Augustine's, indeed, had been a fine choice for such a mind, not only because of their liberal attitude to what was considered non-academic work, but also because they possessed a well equipped laboratory, to which he was given the key. It has often been noted that the nineteenth-century education system, in particular those schools providing an education for middle-class children, was woefully short of scientific training. Wiener (1981) in particular has argued that the typical British public school was more concerned with the liberal arts education which imbued the pupil with a 'distinctive English view' of the world. There is much to be said in support of such an opinion, but it would appear that St Augustine's had adopted a more open approach to its curriculum, and Ferranti undoubtedly benefited from the technical facilities provided at a formative stage in his intellectual development. He had already started to show an embryonic ability to think originally when he was only eleven, informing his parents in March 1876, 'I have made a picture of a water heater and condenser combined which I have invented so that hot water from the condenser will run into the boiler instead of steam coming out of the funnel.' His artistically oriented parents were rather perplexed that he might be overtaxing himself, but he comforted them with the interesting statement that all his ideas 'seemed to come to me naturally', an indication of the thought pattern evolving by his early teenage years. It was in that work room at St Augustine's that a highly original and creative mind was being moulded. At the same time, it is important not to exaggerate the sophistication of the school's facilities or the scientific training provided by the staff because, by the early months of 1880, Ferranti had already advanced at such a pace that the science teacher had been forced to introduce him to a local photographer and amateur scientist, A.J. Jarman, in an attempt to supplement what St Augustine's had to offer. This contact proved to be of crucial significance in channelling Ferranti's mind into a new field, electrical engineering, providing the level of tuition required if practical results were to follow from all the study and tinkering.

It had been during the winter of 1878 that Ferranti had first turned his attention towards electric lighting, after he had been given Ganot's *Treatise on Physics*. This book contains one of the most detailed descriptions of the theory of electro-magnetism available in the 1870s, but unfortunately Ferranti totally misunderstood a section on the properties of magnets, to such an extent, in fact, that he reported to his parents the invention of the Ferranti perpetual motion machine. From his early confusion over electrical theory arose two vital features of his career: firstly, a supreme confidence in his own ability and, secondly, a fundamental faith in the prospects for electricity, the extended use of which he thought would be a 'blessing to the world'. This was a sentiment shared by a growing number of people in the late 1870s, even though electrical technology and technical training in this subject were still in a very early stage of development. Even the terminology for the basic phenomena of electricity had yet to be finalised, and it was not until the International Exposition of Electric Lighting in Paris, in 1881, that agreement was reached on the standardisation of technical nomenclature. This typifies the primitive state of the industry at this time, providing an enormous challenge to the enterprising group of electrical pioneers who were beginning to think more positively about the design of generators and lighting equipment.

Although it had been almost half a century since Michael Faraday had announced his pioneering discovery of electro-magnetic induction in 1831, demonstrating the possibilities of converting mechanical power into electricity, the design of effective and economical generators had made only slow progress by the 1870s. One of the most important discoveries in the intervening period had been made in 1866, when, working independently, W. Siemens in Germany, and H. Wilde and S.A. Varley in England had devised a self-exciting dynamo, dispensing with the need for cumbersome and inefficient permanent magnets. Even so, by the 1870s electric lighting was only being used extensively in lighthouses. Those electrical firms which did exist, for example R.E. Crompton's business, were confined to publicising their wares at exhibitions — Crompton actually described his team as a collection of 'scientific travelling showmen' — and Ferranti frequently attended these events while on vacation with his step-brother, Wladziu, who lived in London. Ferranti saw his first working dynamo when visiting the London offices of the Anglo-American Brush Electric Light Company (Brush) in Hatton Garden, London. This was the most successful company in the early electrical industry, as we shall see later, using an American-designed generator to exploit the growing demand in London. In his description of the event, Ferranti admitted to his parents: 'I went away so delighted that I lost my way twice with thinking of it.' Like many of his contemporaries, he was convinced that Britain was at the dawn of a new 'Electrical Age' and, as Parsons (1939) illustrates, after 1880 an increasing number of installations were being laid out by such firms as Brush, Crompton and Siemens. It was also in 1880 that the independent production of an incandescent lightbulb by J.W. Swan in England and T.A. Edison in the USA was arousing near-panic among gas company share-holders, reflecting the impact electric lighting was beginning to have on the public mind in these years. The gas companies, of course, were worrying needlessly, because, as Byatt (1979) notes, gas remained up to one-third cheaper than electricity for much of the period up to 1914, and the electrical industry was faced with a huge problem in ousting its major competitor from the lighting market.

It is clear that, in moving from mechanical to electrical engineering, Ferranti had latched on to one of the more important technological developments of the era, and it was A. J. Jarman's role to convert his raw energy into practical results after the early stumble with magnetic theory. After the first meeting, in February 1880, they soon progressed from designing an electric bell for the school to copying the small Siemens dynamos which had won the prestigious Trinity House tests for the Lizard lighthouse contract in 1876. This served to channel Ferranti's attentions into the most important sector of the emergent electrical industry, the design of generators, and although it was to be over two years before he patented his own design, this was useful training in the principles and construction of such machines. It was becoming increasingly apparent to Juliana and Cesar that their artistic aspirations were being subordinated by their son's overpowering urge to pursue a career in industry. In fact, as early as January 1879, in what seems a remarkable letter from a fourteen-year-old to his mother, he had mapped out his future. First, he hoped to go to the University of London, which he had read in the *Telegraph* was 'the great place for engineering' and that 'throguh it an extensive field was opened to manufacturing engineers'. He then went on to explain: 'I think the best thing for me will be for me to be will be [sic] a Manufacturing Engineer as then everything I invented I will be able to carry out with advantage.' For one so young to have such definite ideas on his place in life, and such confidence in his ability, is quite amazing, and it would have been difficult for his parents to deny the rationale behind this scheme. Cesar would have preferred his son to have taken the Matriculation Examination, but Ferranti argued that he already possessed the necessary qualifications for his career − an inventive mind, mechanical aptitude, artistic ability and eternal optimism.

The first leg of Ferranti's plan was completed when, in October 1880, he registered for a course in Engineering at University College, London, his father having dutifully paid the fee of thirty guineas. Cesar only finally relinquished his hopes of a Matriculation pass after St Augustine's recommended another year's study. His mother also gave him permission to give up his piano lessons, explaining to one of her daughters that 'his mind is so full of science and inventions and he is longing to be occupied with some construction or another'. He simply had no time for the arts. It is, on the other hand, debatable whether or not Ferranti benefitted from his brief sojourn in the academic world. He was, of course, provided with some of the most advanced laboratories in the country, recently installed by Sir Alexander Kennedy, the man, according to Sanderson (1972), who 'virtually created engineering as a modern university subject in England'. In fact, only the Finsbury Technical College offered anything like the same quality of training as that available at University College, but, as we noted earlier, Ferranti's lack of academic application proved to be a major obstacle in the assimilation of theoretical work. As he later recalled in his speech to the IEE Commemorative Proceedings in 1922: 'I always found it difficult to learn ... I was wanting the whole time to get on to something of a practical nature in the industry for myself.' This again highlights the academic shortcomings of the man, a point to which we shall return in the next chapter when analysing his pioneering work with high-tension voltages.

The opportunity 'to get on to something of a practical nature' came abruptly and tragically, because in February 1881 Cesar fell seriously ill after suffering a cerebral thrombosis, an illness that afflicted his mental capabilities for the rest of his life. In the short term, it meant that a university education was a luxury the family could no longer afford, and Ferranti was obliged to leave University College and find a means of supporting himself and the family. For the first time in his life, he had been wrested from the cocoon of domestic tranquillity. No longer would a short note to his parents procure the necessary working materials, and it would be accurate to describe his efforts over the next few years as motivated as much by the financial insolvency of the family as the desire to make a real contribution to electrical technology. He was, however, by no means overawed by his new responsibilities, informing his father in 1882: 'I know that I have the *strength*, the *ability* and the *determination* to overcome all *difficulties*.' This is a measure of the confidence possessed by the seventeen-year-old, an attribute which always spurred him on in the moments of adversity he was frequently to experience during his career as an electrical engineer. It was also an attribute which, along with his cheerful personality, convinced a series of business associates to provide the backing he required, and in the following chapters it will become clear that this combination of persuasive charm and undoubted technical ability was to see him through a difficult first decade in business.

Leaving University College, of course, would have been only a minor setback to the young engineer, because of his unsuccessful attempts at absorbing university-level work. It also allowed him greater freedom to concentrate on his electrical experiments. By this time, he had progressed from copying other makes of generator to developing his own ideas on the subject, producing what he called his 'Rabbit' direct current generator (dynamo) in the early months of 1881. He was even able to sell the 'Rabbit' to a London dealer in electrical apparatus, Caplatzi in Euston Road, for the princely sum of £5 10*s*, but this was very much a technical *cul-de-sac* for the young engineer and little more was heard of his work on direct current machines. More importantly, it failed to solve the family's pressing financial problems, forcing Ferranti to forgo his practical–experimental work in an attempt to secure a more regular income. He had always been a frequent visitor to the growing number of electrical exhibitions in London, as we noted earlier, and the wide range of contacts he had established were to prove very useful in this situation. One of these contacts, a Mr Grout of the South Kensington Museum, on hearing of Ferranti's dilemma, provided him with a letter of introduction to Alexander Siemens, the head of research at the Siemens company in Woolwich. Ferranti was not actually successful in his first interview, but this did not deter him because he returned the next week and placed an order for carbon rods before Alexander Siemens. As a result of his guile, and obvious enthusiasm, he was taken on as a research assistant, providing the opportunity to enter what was already one of the leading firms in the electrical industry. Siemens were particularly renowned for their avowed interest in research and development, an attribute inherited from their parent firm in Germany and extended by the anglicised Chairman, William (later Sir William) Siemens. This gave Ferranti access to some of the most advanced facilities in the industry, at a time when, as we saw earlier, electrical technology was only in a primitive stage of development.

Although progress in electric lighting had been slow up to 1879, by the summer of 1881 a large number of generators had been installed around the country, and especially in London (Parsons, 1939). The three firms Brush, Siemens and Crompton dominated the industry at this time, accounting for the majority of generators and arc-lights purchased for use largely in isolated installations. As we noted earlier, Brush possessed the most popular designs, the *Electrician* of March 1880 commenting that their generator was the 'best adapted to meet the existing demand, and the larger portion of the lights actually in use in this country are Brush'. By 1881, their annual sales had reached £80,000, according to Byatt (1979), while their nearest rival, Siemens, could only claim £36,000. It must be emphasised, however, that in the early 1880s the manufacturers were only capable of producing a machine which could light no more than thirty-two arc-lights. The era of mass electric lighting using incandescent light-bulbs was very much a phenomenon of the future, even though early in 1882 Edison installed a dynamo capable of serving 1,000 of his new lamps in the first public supply station in this country, at Holborn Viaduct, London. This turned out to be nothing more than a publicity stunt to advertise Edison equipment, but it illustrated the growing sophistication of electrical machinery. It also forced the leading British manufacturers to reassess their own designs, adding to the general feeling that the electrical industry possessed tremendous potential as the growth sector of the future, and, with the emerging market for this new form of lighting, new openings were appearing for men with ambition and foresight.

The young Ferranti was clearly eager to join this group of electrical pioneers, and in June 1881 he started work at Siemens as assistant to the firm's Chief Research Engineer, Dr E. Obach. At that time the company was developing a new range of dynamos in response to the factors outlined earlier, placing Ferranti at the very heart of the industry. As he reported to his mother, the job 'is as good for me as if I was spending piles of money weekly on experiments', giving him first-hand experience in the design and production of new machinery. He seems to have impressed his employers quickly with his grasp of the subject because by September he had been placed in the much more responsible position of installation engineer, travelling to the Wolverhampton Ironmasters' Exchange and the Crown Point Printing Works in Leeds to fit electrical generators. At a time when electricity was struggling to break into the lighting market against the fierce competition of its cheaper counterpart, gas, it reflects great credit on this seventeen-year-old that Siemens were willing to entrust their reputation to him. Such was their confidence in his ability, indeed, that he was invited to assist in exhibiting the Siemens products at the prestigious Smoke Abatement Exhibition held in London from November 1881 to February 1882. It was there that Siemens generators showed to a curious audience that electric arc-lights did not vitiate the air in the same way that gas lamps produced their glare.

He was even fortunate enough to work with William Siemens on the electric arc furnace project, carrying out a series of experiments with one of the established leaders of the electrical industry and later demonstrating the test model at Owen's College, Manchester. This illustrates just how crucial this first post proved to Ferranti's development as an engineer, providing the kind of apprenticeship which existed hardly anywhere else in the country.

In the early 1880s the electric light was beginning to make rapid strides, and the growing interest in the new technology was not only attracting the attention of established companies like Siemens, Brush and Crompton, it was also offering opportunities to individuals with a keen eye for an opening. While working at the Wolverhampton Iron-masters' Exchange in September 1881, Ferranti had met a Mr Alfred Thompson, who was an associate of Cesar's in the photographic business and an electrical enthusiast. Thompson was interested initially in the Siemens dynamo Ferranti was installing, but in conversation the older man became captivated by the ideas and schemes his new friend discussed. Such was the impression Ferranti made on Thompson that the latter decided to introduce him to a lawyer, Francis Ince, another amateur scientist. This meeting, recorded in detail by de Ferranti and Ince (1932), proved to be a vital watershed. The two men agreed wholeheartedly that Ferranti was wasting his time working for Siemens, and Ince in particular was concerned that the company would 'rob your brains. You'll do the inventing and they'll collect the cash.' Ferranti naïvely asked if they thought he should ask for a rise, but Ince strongly advised against it and instead proposed: 'You must start right away on your own.' Here was the break for which the young Ferranti had been waiting in order to fulfil his avowed intention of becoming a 'Manufacturing Engineer'.

The real reason why Ince and Thompson took such an interest in Ferranti, apart from his obvious enthusiasm and confidence, was the plan already formulated by the young engineer to improve on the form of armature used in the Siemens alternator. This innovation was currently taking shape in the workshop owned by his earlier guide and mentor, A. J. Jarman, who had moved to London in 1881 to expand his photogrpahic business. Such was the perceived potential in Ferranti's idea that, after several more meetings early in 1882, by July Ince and Thompson had agreed to provide him with a salary of £2 per week, plus all working expenses, while the machine was being developed. This naturally prompted Ferranti to leave Siemens on 12 July, a turn of events with which the company was less than pleased. There followed an acrimonious exchange of letters in the *Times*, in which the validity of both the idea and Ferranti's business methods were questioned, but no legal challenge ensued and the development work continued. Evidently, preparations had been well in hand, for on 18 July the provisional specification for a generator was lodged with the Patent Office by the patent agents who were to serve Ferranti for many years to come, Marks & Clerks, and over the next three months he worked to perfect his machine to the satisfaction of his new partners.

By the summer of 1882, then, Ferranti had already come under the influence of the man who was to be his guiding hand over the next eighteen years, Francis Ince. Ince was a successful solicitor specialising in company law, having first entered the business in partnership with J. P. Ingledew in 1865. They had soon expanded from their original base in Cardiff into larger premises in London. By the 1880s, the firm, now called Ingledew Ince & Colt, of Benet Chambers in Fenchurch Street, was one of the most prominent in the coal and shipping industries, counting among its clients the Coal Depot Owners' Association, the Falkland Islands Company, and, reflecting their links with the new industry, the Hammond Electric Light & Power Supply Company. Ince was a man of many interests who liked to tax his mind and drive himself to the limit, being a far-sighted optimist, yet a person who insisted upon orderliness and businesslike behaviour in all aspects of life. Ferranti could hardly have found a better business partner, because Ince provided that counter-balance between his young associate's desire to improve the product until it was perfect and the need to sell at a profit. Later passages will describe the often fiery relationship between the two; Ferranti was expansive and eager to improve, while Ince was a firm believer in natural growth through sound business practices. It was a relationship not unlike that between Hugo Hirst and Gustav Byng in the early history of GEC, as Whyte (1930) relates, another marriage of the ambitious and the conservative, but it was one which proved successful in providing balanced leadership in a period characterised by severe market difficulties in the electrical industry.

It would appear that Ferranti had again been fortunate in his choice of associates, but it is just as clear that he had produced his idea at a most opportune time. Although it was to take many decades before the electric light lost the label of 'luxury light' applied to it by William Siemens in 1882, the development of incandescent light-bulbs and more powerful generators was stimulating ever-wider interest in this new method of lighting. The public exhibitions had also been increasing in frequency, and in January 1881 the *Electrician* was able to report that 'there has been great progress in every stage of electric lighting: apparatus has been improved, extension has been great, and public opinion has been educated'. In that year, as we saw earlier, Brush sales of electric lighting apparatus totalled £80,000, but in 1882 they soared to over £200,000, Siemens' sales rose from £35,800 to £83,780, respectively, and Crompton was clearly busy. Unfortunately, though, this feeling of anticipation manifested itself in financial dealings of a highly dubious nature on the Stock Exchange. The new illuminant, of which so many spoke in over-optimistic tones, proved too great a lure for the less scrupulous company promoters, all too many of whom were anxious to capitalise on the ignorance of the non-technical investor. The Brush Company was the chief culprit, floating fourteen subsidiaries which the *Statist* reported monopolised the attention of financiers by their outlandish claims concerning the potential demand for electric lighting and the supposition that it could be sold at the same price as gas (Byatt, 1979). In fact, this investment mania is known as the 'Brush Bubble', during which Cottrell has calculated that over £7,000,000 was raised by the electrical industry. Clearly, on the basis of this evidence, investors were not averse to risking capital in new projects, but, at the same time, the Brush Bubble illustrates the almost complete state of ignorance exhibited on the floor of the Stock Exchange about industrial securities.

The Brush Bubble of 1882 is remarkable in highlighting how easy it proved in boom conditions to persuade investors that a quick return could be made from such speculative ventures. Had the financiers investigated the claims made by Brush in particular, that electricity and gas could be sold at comparable prices, they would soon have discovered how the company was blinding them to the harsh realities. Even using the new incandescent lamps, electricity still remained up to one-third more expensive than gas, as Byatt (1979) reveals, while most of the Brush concessions were for the much less efficient arc-lighting system. It also transpired that only commercial customers would purchase electric light, and the bulk market failed to materialise for several decades in Britain. Despite these negative indicators, the Brush £10 share rose to a peak value of £68 in May 1882, but it soon became apparent that the promises could never be fulfilled, and the fourteen Brush subsidiaries started making losses. In fact, at the January 1883 Brush Annual General Meeting the Chairman was obliged to resign, after accusations of bad management, and a general feeling of mistrust descended over the electrical industry within financial cirlces. It would appear that a large proportion of the £7,000,000 raised during the Brush Bubble had either been taken in company promoters' fees or was written off as a bad investment by the end of the year as the losses started to appear in electrical company balance sheets. Little had actually been invested in the industry proper, largely because the projected level of demand had never been achieved, casting a dark shadow over electrical investments in British financial circles. 1882 marks the beginning of the electrical industry's poor relationship with British financiers and, as we shall see later, the bitter experiences of that summer remained in their memories for many years to come. Fortunately, though, Ferranti was not to suffer from any backlash for two years and, indeed, was able to raise funds after the Brush Bubble had burst.

Obviously, the £2 per week provided by Ince and Thompson in July 1882 was more modest than the scale of investment others had engaged upon, but it is apparent that Ferranti was placed in the ideal climate in which to canvass financial support. Ince and Thompson must also have realised this and, anticipating the need for larger amounts of working capital, they discussed the market potential of Ferranti's machine with some of the most eminent names in the industry. We have already noted that Francis Ince had a direct link with the Hammond Electric Light & Power Supply Company, and this contact he used very well. On 14 September, according to Juliana's diary, Ferranti, Ince and Thompson 'went to meet some gentlemen in the City' and it is surmised that among this group was Robert Hammond, one of the foremost electrical pioneers of the 1880s and the man who was to be the sole licensee of the Ferranti alternator. It became clear that there would be sufficient financial support from Hammond for the proposed venture, and on 15 September the firm of Ferranti Thompson & Ince Limited was created with a nominal capital of £250,000, composed of 2,000 £5 A shares and 36,000 £5 B shares. This established the Ferranti name in the market for the first time at the dawn of what many saw, incorrectly, as the 'Electrical Age'.

Until 1881, Robert Hammond had been in partnership with his

brother in the iron trade, when he visited London to inquire into the best method of lighting an iron works. Recognising the merits of the Brush generator, and foreseeing a wide demand for arc-lighting in manufacturing works, he contracted with the Brush Company, in July 1881, to sell their products in the north of England. Hammond soon established a large business and, by the time of the Brush Bubble, he was regarded as one of the most successful salesmen in the industry. This led him into closer contact with Brush, and in 1882 he was appointed Managing Director of several of the newly-floated Brush subsidiaries. Hammond recognised, however, that, if electric lighting was to compete with gas, a system of bulk supply was required, but the Brush Company were reluctant to produce machines with a capacity greater than 200 lights, and the Ferranti machine promised better results. At the same time, Brush were by then beginning to experience the backlash from the share mania, and Hammond was concerned about the company's shaky financial base. Although he had signed a long-term contract with Brush and was obliged to continue the agency agreement, his decision to invest in Ferranti Thompson & Ince Limited showed astute judgement, because the industry was showing great interest in their new machine. The *Statist* reported in September 1882 that 'Brush shares have suffered from the attention given to a new economical dynamo ... of which the Hammond Company have obtained the right of use.' Ferranti Thompson and Ince Limited had been able to sell all 2,000 of their A shares to ninety-three separate shareholders, while retaining all the B shares as reserve for future growth, but the key figure, of course, was Robert Hammond, who purchased 1,875 shares and paid a £5,000 deposit for the licence, in anticipation of a market for more powerful generators. The Ferranti alternator was coming to be regarded as one of the few capable of providing sufficient power for bulk supply schemes, although, superficially, in the use of two crowns of fixed coils with iron cores, he had not departed significantly from standard practice. His principal digression had been in the design of a novel form of armature: working along the general principle that the armature must be as light as possible, in order that it might revolve at high speeds to produce more current, Ferranti developed an armature of copper strip only one inch wide, moulded into the requisite zig-zag shape. This resulted in the lightest form of armature available, as well as one capable of running within the most powerful area of the magnetic field, making it more efficient. It was this power and efficiency which made it so suitable for the bulk supply schemes envisaged by Hammond, but only after a major hurdle had been negotiated could manufacture commence.

Unfortunately, Ferranti had not been the first to propose the use of this form of armature because, unbeknown to him, Sir William Thomson (later Lord Kelvin) had taken out a patent a year earlier for just such an idea. It is perhaps to the credit of the young Ferranti that he had been working along the same lines as one of the most eminent scientists of the nineteenth century, but Juliana describes in a letter to one of her sisters 'a great trouble' in August when the situation was discovered. Immediately, the three partners travelled to Glasgow and a deal was negotiated, whereby Sir William was to be paid £500 per annum for the use of both the patent and his name. This proved to be a big fillip to the young company because, in such a fickle market, having Sir William Thomson's name attached to the product — it was marketed as the Ferranti-Thomson alternator — added a welcome touch of credibility. The agreement also provided Ferranti with an important source of advice because during the 1880s Sir William frequently corresponded with him on all aspects of electrical engineering. Like Jarman, Grout, Siemens, Thompson and Ince before him, Sir William was impressed with this dark-haired, eager youth who exuded such confidence in his own work, and the coincidental patenting of the 'zig-zag' armature proved to be a twist of fate that had several short- and long-term benefits.

The big test for the Ferranti—Thomson alternator came at the first public trial in November, when the technical press were allowed to make an exacting study of the machine. It would appear that things went well, because there were few adverse criticisms. In fact, after testing Swan incandescent lamps on two circuits powered by Siemens and Ferranti generators, the *Electrician* reported 'startling results': the former could light eight lamps per horse power, the latter 12·4. The *Engineer* concluded that the Ferranti—Thomson machine could certainly do more work per horse power 'than any other good machine already in the market', providing just the kind of publicity Hammond required for his plan to create a market for bulk distribution systems. Technical acclaim, however, would appear to have been no guarantee of commercial success in the harsh climate of the 1880s. 1882 was very much a false start in the history of the British electrical industry, and it was only in the next decade that real growth was to occur. Although Ferranti had at least been able to achieve his ambition of becoming a 'Manufacturing Engineer', and a small workshop in Appold Street, London, was rented as his first premises, the new enterprise was to go through a baptism of fire over the next three years as the industry lurched slowly out of the euphoria of the early 1880s.

It was fortunate for Ferranti Thompson & Ince Limited that they had taken full advantage of the situation in 1882, for, by November 1883, the *Electrician* was reporting that 'the want of confidence' in the industry was a major reason for 'a year of promise unfulfilled'. Sir Charles Young, the Hammond Chairman, also noted that, because of the many different systems on offer at that time, many customers were reluctant to commit themselves until one had attained dominance. The 'Battle of the Systems', as it came to be known, between the Alternating Current (AC) and the Direct Curent (DC) camps belongs to the next chapter, but it must be noted that, at this early date, Ferranti had already placed his faith in the system that was to become standard by the end of his life. He was convinced that, if the electrical industry was to make deep inroads into the lighting, heating and power markets, the AC system would offer the cheapest methods of generation and transmission. It was already clear that AC gave the electrician the opportunity to transmit at higher voltages (pressure), thereby reducing the amount of copper required in the cable and decreasing its price. On the other hand, the DC system, because it could only use low voltages, required a good conductor in order to prevent the loss of pressure, consequently making the cable costly. With this advantage in distribution costs, the AC station could be situated in isolated areas where land was cheap, water available and fuel easier to transport, while the DC stations were always built within their markets to guarantee full delivery of power. The problem facing Ferranti and other engineers like W. Mordey who advocated the use of AC was that the system had yet to be fully developed, and the choice of an alternator went against a considerable body of authoritative opinion which doubted whether bulk distribution systems would prove feasible. Later chapters will demonstrate the long-term wisdom of Ferranti's choice of system, but in the early 1880s not only was he fighting an uphill struggle with the technology, he and Hammond also faced severe difficulties selling bulk distribution systems in a market dominated by the demand for isolated installations. The situation was made even more uncertain by the intervention of the State in its attempt to control public utilities, compounding the feelings of unease already evident after 1882.

As early as 1878, a Select Committee chaired by Sir Lyon Playfair had investigated the need for controls over this new industry, but at that time intervention was not thought necessary because of the technology's experimental nature. It was only after the burst of activity in 1882 that Parliament decided to introduce statutory controls, bringing in legislation which was to influence the development of electricity supply for the next half-century. The government's policy was heavily determined by the President of the Board of Trade, Joseph Chamberlain, a leading advocate of the 'Municipal Socialism' movement which insisted on greater municipal responsibility for providing and controlling public utilities. There was also general concern that gas and water companies had, in the past, pursued profiteering strategies, and after the 1870 Tramways Act public utilities were to come under much closer public scrutiny. It was in this context that the 1882 Electric Lighting Act was passed, and three clauses deserve special mention: firstly, a private company had to seek permission from a municipality before the Board of Trade could sanction a provisional order to light a prescribed area; secondly, the private company's lease only lasted for twenty-one years, after which the municipality could purchase the system without providing compensation for goodwill or loss of earnings and, thirdly, no systems were allowed to supply electricity over more than one municipality. Contemporaries like Emile Garcke (later the founder of British Electric Traction) castigated the government for imposing such restrictions on private enterprise, but, as both Hannah (1979) and Byatt (1976) argue, it is difficult to find hard evidence to support such a view in the 1880s. One could never dispute, for example, the proposition that twenty-one years was sufficient time to make a return on the investment required to build a generating plant in this decade, while only in the 1890s did electrical engineers begin to construct inter-municipal networks. The most important aspect of this legislation was the power vested in the local authorities, and over the following five decades this feature of the electricity supply industry was to create major problems for the manufacturer, as we shall see in Chapters 3 and 5. In the early 1880s, on the other hand, the legal measures simply acted as a psychological deterrent to electrical activity because the more important obstacles to progress remained the financial and general economic conditions prevailing after the collapse of the Brush Bubble.

Clearly, by the summer of 1883, market conditions had changed alarmingly after the euphoria of 1882. The electrical industry was not alone in suffering from a collapse in demand, of course, because the mid-1880s were to bring problems throughout the British economy, but as a new industry it had few resources on which to fall back. The financial sector was not to be induced into a repetition of the 1882 mania, and the gloomy prospects were forcing severe cutbacks in the major electrical companies. In the light of these changed circumstances, Hammond decided to rationalise his business interests by consolidating the assets and executive administration of his Brush subsidiaries into his own company. Ince and Ferranti were persuaded that this process of horizontal amalgamation was the only logical move and, by November 1883, Ferranti Thompson & Ince Limited had been voluntarily liquidated and a new company, the Ferranti Hammond Electric Light Company, had been created to acquire both enterprises. This allowed Ferranti and Hammond the opportunity to pursue a more integrated approach to the market, publishing catalogues to advertise their ideas for bulk distribution systems and attempting to standardise the range of machines on offer to potential customers. Ferranti took on the role of Hammond's technical consultant, escorting him on visits in the desperate attempt to convince local councillors that electric lighting had much to offer. In the meantime, Ince took a back seat in the company, advising only on legal matters, while Thompson would appear to have left the electrical industry to concentrate once more on his photography.

Although the Hammond and Ferranti companies had been merged in November 1883 as an economy measure, this was not accompanied by a contraction in the size of the new business because, by the end of the year, two new branches had been added to complement their aim of supplying bulk distribution systems. We have already noted the scarcity of technical education in the 1880s, and Robert Hammond was so concerned about this issue that he created the Hammond Electrical Engineering College. He told his shareholders in 1884 that it had been 'first started to supply a want which was felt throughout the electrical world. There were no skilled electricians combining practice with theory.' Run by H. E. Harrison, the courses were divided into a first year of theoretical training and a second year of practical work in the Appold Street manufacturing premises, providing the firm with a supply of trained engineers. It also reduced the need for heavy expenditure on training because the students paid their own fees.

After the collapse of the Ferranti Hammond Company in 1885, this was the only part of the organisation to survive intact and, although it was renamed Faraday House in 1887, the college continued as one of the most prestigious training institutions in the industry up to the 1960s.

The second branch added to the company, the manufacture of incandescent light-bulbs, proved to be a much more hazardous venture, particularly in view of the strong hold exercised by Edison and Swan over this market. In 1883 these pioneers, although never actually meeting personally, combined their technical expertise in Britain by creating the Edison & Swan United Electric Light Company (EdiSwan). J. S. Forbes, a redoubtable businessman of the era whom we shall meet again in the next chapter, was appointed Chairman of this enterprise and, under his leadership, the EdiSwan product came to be the natural market leader. To have challenged this position in a contracting trade would have been foolhardy, but Hammond acquired the rights to manufacture the Wright & Mackie bulb and purchased a factory in Bermondsey Street, London, where he hoped to establish a successful rival to EdiSwan. Ferranti was closely associated with this project, patenting a method of manufacturing the bulb in 1883 and designing the layout of the factory, but it was almost a year before the first products were ready for the market and the whole venture proved to be a serious drain on resources. They even tried to break the EdiSwan hold by selling at 3*s* 6*d*, 1*s* 6*d* cheaper than the standard London price, but there is no doubt that the bulb factory was a commercial disaster, compounding the general problems experienced by the Ferranti Hammond Company after 1883 in the generator market.

The key to the success of the whole strategy was, of course, the construction of bulk distribution systems using the Ferranti alternator but, despite vigorous salesmanship on the part of both Hammond and Ferranti, not a single contract was forthcoming. This was not unusual for the industry at this time, because even by 1888 only one electricity supply company was operating in Britain (in Brighton), ironically established by Hammond in 1881 using Brush equipment. Hammond did succeed in securing a provisional order to light Hampstead, but for this he required financial support which was never to materialise. As he reported in 1885: 'Every shareholder gave the same answer, that they had lost enough money by electric lighting, that they had been too soon in it, and did not care to put more money in it ... There was none who would risk £100 in it.' W. P. Kennedy (1974) has criticised British financiers for their 'risk-aversionist' tendencies in the decades leading up to the First World War but, in the light of the huge losses suffered in 1882, the reluctance to support Hammond's venture appears to have been highly rational. Indeed, in the mid-1880s it was difficult to raise capital for any industrial ventures, such was the low level of activity within the economy as a whole, but the soured relationship between electrical companies and financiers was to have longer-term implications. For Ferranti Hammond and Company such inanition proved disastrous in the short-term and, as we shall see in Chapter 3, British electrical companies continued to suffer liquidity problems right up to 1914, largely as a result of the spurious activities of 1882.

Ferranti Hammond's failure to secure any orders for large-scale projects seriously undermined the rationale behind the whole strategy. The company was able to sell several alternators for use as isolated installations, for example to the First Avenue Hotel in Holborn and Peterhouse College, Cambridge, but it is doubtful whether even these contracts made much profit. The problem here was the one-off nature of each order, which prevented the standardisation of design and production, while Ferranti's insistence on absolute perfection before delivery further compounded the situation. It is clear that Ferranti found it difficult to make the transition from laboratory assistant at Siemens to a manufacturer of commercially-viable products, and in May 1883 Francis Ince was obliged to close down the Appold Street works temporarily in an attempt to instill greater discipline into his young partner. Ince had written to Ferranti in April warning of the consequences of uneconomic production methods: pointing to the large number of prototypes designed but never actually completed. He went on:

It is of great importance that no machines should ever stand for improvement and if you insist on this we shall be improved off the face of the earth, and others with a less perfect machine will do all the work and laugh at us while we are theorising ...

Constant tinkering and alterations in the machines now in hand simply means *great expense* in finishing them; and a future sale *no one knows when*, without an atom of profit.

It was an object lesson in production management, but it is equally clear that Ferranti failed to heed such lessons and the temporary closure in May was the only way in which Ince could restore sound business practice in the factory.

These were the harsh realities faced by Ferranti in the mid-1880s, but an even greater shock to his career lay in store for him. With the severe cutback in electrical activity after 1882 and the failure to sell any bulk distribution systems, Ferranti Hammond's financial commitments in terms of production and development were just too heavy for the flow of business. In 1884, when the company reported a loss of over £30,000, severe cutbacks in expenditure on experimentation and advertising were imposed, giving Ferranti another harsh lesson in business economics, while in the ensuing twelve months prospects deteriorated even further. In the seller's market of 1882, Hammond had been obliged to sign a long-term agency agreement with Brush, and generators worth over £10,000 had also been stockpiled. The company had fought vigorously against the vagaries of the market and the inanition of financiers, but in June 1885 it was finally decided to liquidate Ferranti Hammond and bring to a temporary halt their careers as electrical engineers. The company's major rivals, Siemens, Brush and Crompton, had also experienced hard times and, although they survived, it is apparent that the early 1880s was not providing an environment conducive to success in the electrical industry. Abroad, as we shall see over the next two chapters, strong electrical industries were emerging, but in Britain the industry remained in a state of suspended animation for much of the decade, and Ferranti, as much as anybody, was badly affected by these conditions.

This severe baptism of fire for Ferranti must have come as a great shock after the successful launch in 1882, but one of his strongest characteristics was always an indomitable spirit. His confidence in his own ability and his refusal to be defeated by any problem, whether commercial or technical, were assets to which we have already referred, and in 1885 they were very much to the fore. With sketch-books brimming with ideas, inventions and schemes, he was able to persuade Francis Ince to join him in a new partnership, sure in the knowledge that his technical abilities were capable of producing more encouraging results. Nowhere is Ferranti's enterprising spirit better illustrated than in this new venture, for, within a month of Hammond's collapse, he had installed some machinery in his new premises, 57B Hatton Gardens, London, and with Ince had begun trading under the name 'S. Z. de Ferranti'. A third partner, Charles P. Sparks, also joined the firm in 1885, beginning what was to become an illustrious career in the electrical industry, culminating in his election as President of the IEE in 1915. Sparks had been one of the more promising students at the Hammond Electrical Engineering College and, over the next fifteen years, was to act as Ferranti's right-hand man, translating patents and rough drawings into saleable products. Ferranti had obviously failed to grasp the need for commercial manufacturing techniques by 1885, and in this context 'S. Z. de Ferranti' appears to have been a highly functional partnership, with Sparks providing that link between technical excellence and commercial reality, while Ince supervised the financial management of the business. On this more prudent basis, Ferranti was to enjoy much greater success as a 'Manufacturing Engineer' after the traumatic start to his career.

By 1885, after his exploits with Hammond, Ferranti had already established himself as an important member of the British electrical fraternity, a significant achievement for a 21-year-old. He had been provided with a good deal of support, not only from his family up to the Easter of 1881, but also from a number of experienced businessmen and scientists. What is evident, however, is that Ferranti was largely self-educated. His parents' artistic abilities, and stories of the achievements of the Ziani family, must have contributed to his thinking, but it was through practical interests, personal study and contacts with knowledgeable people that he was able to develop his capabilities. Perhaps it was his Italian blood which aroused such passion, but the confidence and enthusiasm expressed about both his own abilities and the potential of electricity persuaded others to provide the necessary capital for his ideas. The creation of 'S. Z. de Ferranti' stands out as a classic illustration of how this commitment to electrical technology was driving him on to bolder deeds, and we shall see in the next chapter how he became a central figure in the industry by forging ahead with his own ideas. But why the exceptional, and in some ways naïve, motivation? The answer lay in his intuitive ability and the desire to change the world by the application of technology. This was the driving force, not just to dream but to succeed in making reality of the dreams, highlighting the absorbing and eventful nature of Ferranti's story.

2

The Deptford experience

One of the most significant episodes in the early history of the British electricity supply industry is Ferranti's attempt to build the first modern power station at Deptford and establish high-tension AC as the principal form of electricity transmission. This, according to Parsons (1939), must be regarded as 'one of the boldest conceptions in the whole history of central station work' and has rightly been seen as a prophetic vision of present-day practice. On the other hand, Ferranti has been criticised by Hughes (1963) for 'overloading an entire engineering-human complex' and for pushing the frontiers of electrical technology too far for the current level of knowledge. In a more recent publication (1983), he has reiterated this argument, asserting that the enterprise of Ferranti and his financial backers was misplaced. It is perhaps ironic, however, that such criticism should be levelled at a member of the British electrical fraternity when its general level of enterprise has also been devalued by Professor Hughes (1962), particularly when compared to their American counterparts. In examining this crucial technological development, we shall test the validity of such arguments, while examining some of the constraining factors holding up the emergence of a viable British electrical industry after the problems of the early 1880s. There were still major problems to be overcome, and Ferranti's ambitious efforts were continually coming up against various obstacles. Once again, we shall see his ability to raise the necessary support for his ideas at a time when others were still suffering from the backlash experienced from the Brush Bubble, but the questions remain whether he used the capital in the most effective manner and whether Deptford made an impact on the development of the electricity supply industry in Britain. Clearly, Ferranti's reputation in this period rested on the success of this venture, and his personal qualities were severely tested in the attempt to push the boundaries of technology forward at what many saw as an alarming rate.

The creation of the new partnership 'S. Z. de Ferranti' was, as we have already noted, a clear illustration of Ferranti's obstinate refusal to give in to the market problems of the mid-1880s. It provided the necessary vehicle for extending both his influence upon electrical technology and his reputation as an electrical engineer. To date, his reputation had been based on the zig-zag armature devised in 1882, but from 1885 it was the development of the less spectacular meter on which he was to base his commercial fortunes. In fact, Francis Ince had only been persuaded to join the new venture after Ferranti had sold the rights to market his novel design for a meter in Belgium and France to the Compagnie Générale d'Electricité (CGE) of Antwerp. This provided the initial incentive to continue in business after the demise of Hammond's plans. In general, the meter had received little attention from electrical engineers, pioneering attempts having ranged from a clock mechanism to measure the length of time electricity had passed through the circuit to Edison's electrolytic meter, but Ferranti was convinced that a more suitable device could be produced. It was also becoming increasingly obvious to a public used to more accurate gas meters that the electrical industry could never be completely trusted until reasonable methods of measuring consumption had been introduced, further evidence of the credibility gap that had emerged in the early 1880s.

During the latter months of 1884 Ferranti had been experimenting with a novel concept in measuring the flow of electricity, using a bath of mercury in which a fan was placed to convert the flow of electro-magnetic pulses into a mechanical movement which activated the read-out mechanism. It is interesting to note that it was from his childhood 'bible', Ganot's *Treatise on Physics*, that the germ of the Ferranti meter was born, proving that his grasp of electro-magnetic theory had improved significantly since his early stumblings with a perpetual motion machine in 1879. This use of electro-magnetic theory was eventually to provide a real breakthrough in meter design, although work was at first thwarted by the failure to immerse the fan completely in the mercury. It was here that the advice of Sir William Thomson proved crucial. Sir William explained that the surface tension of the mercury prevented the fan from turning, and only its complete immersion would lead to the successful development of what Ferranti claimed in his first advertisement of July 1885 to be 'the only commercial meter up to the present time', the mercury-motor meter. Ferranti later admitted in his 1895 address on meter design to the Royal Scottish Society of Arts that these early products were 'at first very incorrect, did not give at all proportionate readings, and would therefore have been of no use from a commercial point of view', but improvements were introduced and by the end of the year loads as low as 0·3 amperes could be measured. This was an accuracy unrivalled by contemporaries and several awards were bestowed on the Ferranti meter, including a Silver Medal at the International Inventions Exhibition of 1885 and a Diploma of Honour at the Antwerp Exhibition. In August 1892, the *Electrician* tested two of these meters, which had been in service for the Kensington Court Supply Company since 1885, and reported that the wear and tear on the mechanism was negligible and the constancy 'by no means unsatisfactory when we take into account the date of the meters; 1885 is almost in the dark ages as far as electricity meters are concerned'. Clearly, Ferranti had developed a meter capable of commercial use, improving the industry's credibility amongst its potential customers and opening up a whole new market for his proposed business venture.

1885, indeed, stands out as a vitally important year for Ferranti, not only because it marks the beginning of his long and successful association with the meter, but also because of the contacts he made as a result of this development. Ferranti's association with the meter was to prove fundamental to the survival and growth of his business in the decades ahead, and in the next chapter we shall take the story up again, but in the short term it persuaded Francis Ince to partner him in 'S.Z. de Ferranti' after the CGE had purchased a licence to sell the meter in Belgium and France. This licence was worth a total of £5,000 over the next three years and, although the electricity market had not grown to any considerable size in Britain, Ferranti was also able to sell his meters to various private installations. The most important of his home sales was to Sir Coutts Lindsay & Company in September 1885, effecting an introduction to one of the most enterprising electricity supply operations in the country. Sir Coutts Lindsay & Company had only recently been established, after Sir Coutts himself had aroused great interest in the lighting of his Grosvenor Gallery by electricity, so much so that many of his neighbours in Bond Street were requesting a supply for both commercial and domestic purposes. This was eventually to result in a most remarkable episode in the industry's history, putting Ferranti on the centre-stage in a period of rapid progress in the technology, and bringing together an enterprising group of electrical enthusiasts. In the meantime, though, he was forced to play a waiting game as others experimented with various methods of supplying electricity.

The Grosvenor Gallery had started life in 1878 as a sanctuary for Pre-Raphaelite painters, under the sponsorship of the wealthy Sir Coutts Lindsay. This man was a member of the younger branch of the Lindsay family, whose head was the Earl of Crawford, a prominent amateur scientist who, according to Appleyard (1939), had been instrumental in founding the Society of Telegraph Engineers, the forerunner of the IEE. It was on the Earl's recommendation that Sir Coutts first installed an electrical generator in the Gallery's backyard to enhance the view of his exhibits, but after several neighbours requested a supply he embarked on the much more ambitious plan of providing a public source of supply. This would ordinarily have involved him in detailed negotiations with his local authority for permission to lay cables in the street but, in an attempt to avoid the limitations of the 1882 Act, Sir Coutts decided on the ingenious, yet technically precarious, ruse of employing high-tension AC and hanging the cables from the roofs of his customers. Had he chosen DC, the weight of the cable would have precluded an extensive system and, foreseeing a large demand for electric lighting among West End clubs, theatres and wealthy residents, he decided on the use of the lighter AC cables. This was a daring decision because, although a Siemens alternator could be acquired to generate at the relatively high voltage of 1,200 V, this pressure was too dangerous for domestic use and consequently each house had to be fitted with a transformer to reduce it down to 100 V. The transformers initially used were of the Gaulard & Gibbs 'secondary generator' type, and it was to be the development of this device that prevented the greater extension of the network in its early days.

Up until 1883, although Kennedy, Jablochkoff, Fuller and Gordon had each designed theoretical systems of distribution using transformers, little expertise in the working and efficiency of this device was available, and Fleming (1930) has noted that AC systems were severely handicapped by this obstacle. In that year, the Frenchman Lucien Gaulard and his wealthy English backer, John Gibbs, installed an experimental fifteen-mile AC network along part of the line owned by the Metropolitan Railway Company using what they called 'secondary generators' for transformers. A year later, at the Turin Electrical Exhibition, they won a prize for a similar system, and by 1885 had secured contracts for their secondary generators in Aschersleben, Rome, Tours and in the Grosvenor Gallery network. The first trial of the permanent Grosvenor Gallery station took place in March 1885, but, as with all their other contracts, the secondary generators failed to come up to expectations. Gaulard had made two basic errors when designing his system: he had built the secondary generators with an 'open' magnetic circuit, giving them a prejudicial electrostatic capacity, and he had placed them in series along the circuit, preventing the use of automatic regulation. In practical terms, this resulted in some major faults which were described in detail by Charles Sparks when assessing the Grosvenor Gallery network in 1886:

1) No two secondary generators gave the same electromotive force.
2) Regulation by hand resistance was so bad that lights were jumping, and many broken.
3) [the consumer] was forced to burn all the lights or none at all.
4) Poor insulation between primary and secondary circuits.

The early electrical engineers could have done without this adverse publicity, and such were the problems experienced by Sir Coutts Lindsay & Company that within six months the station had been closed owing to mounting complaints from customers. Despite assistance from the eminent consulting engineer, Dr John Hopkinson, by the end of 1885 general doubt was being expressed concerning the wisdom of Sir Coutts Lindsay's choice of AC, adding to the misgivings voiced about this method of electricity generation in 1882.

Ferranti was aware of these difficulties, and of the development of a more suitable transformer by three Hungarian engineers, Zipernowski, Déri and Bláthy. Quite characteristically, Ferranti had commented to an acquaintance that 'the mechanical construction of both these types of transformers was very poor and that a more satisfactory one could be devised'. This acquaintance was G. L. Addenbrook, then chief test engineer of the United Telephone Company, a firm which was experiencing some problems in their speech transformers. Addenbrook, through Francis Ince, had sought Ferranti's advice on his company's problem and, although the young engineer had very little knowledge of such devices, within a day of meeting the telephone engineer he had solved the speech transformer problem, using some florists' wire and a shunt coil from one of his meters. Addenbrook later wrote to Ferranti's wife that he was amazed at the rate of progress achieved with such rudimentary materials, but, more remarkably, within five days of patenting the speech transformer in November 1885 Ferranti had also taken out a patent for what Fleming (1930) has described as the first transformer of modern design. This product came to be regarded as a considerable improvement on contemporary designs, in terms of construction, insulation and efficiency, pushing Ferranti into another field of electrical engineering for which his company was later to establish a world-wide reputation in the industry. To complete a fortnight of intensive work, Ferranti then patented a system of high tension AC distribution which Bailey (1932) has called 'remarkable as a prophetic vision of modern-day practice'. It incorporated four features that were to become fundamental: the stepping-up of the voltage at the power station by transformers; the use of tabular concentric mains; the use of automatic control at the sub-station, and cut-out devices to be activated as the demand falls away. This was the master patent to which he worked over the next seven years, but first of all he required a market for the system.

Having established a contact with the Grosvenor Gallery in September 1885 when he sold them thirty meters, Ferranti must have remained in contact and when the beleaguered station was closed down again at the end of the year he was brave enough to offer his own solutions. The directors of Sir Coutts Lindsay & Company were evidently impressed with this young man's confidence because, after dining with them, his mother noted in her diary that 'things seem to promise well for him'. Indeed they did, and more meetings with the more technically aware Earl of Crawford, and a demonstration of the Ferranti transformer in January 1886, sealed the deal. Ferranti was actually appointed Chief Engineer to the station at a salary of £500 per year, providing his business with an excellent shop-window for what was advertised as the 'Ferranti System'. This could not have come at a better time for 'S. Z. de Ferranti' because, although the meter business was beginning to expand, stagnation in the electrical industry had brought the production of generators to a standstill and the Hatton Gardens works were badly in need of business. As far as Gaulard & Gibbs were concerned, Ferranti's appointment signalled the beginning of lengthy litigation over the legitimacy of his transformer but, in a skilful manoeuvre, Ince successfully petitioned for

for the revocation of the Frenchman's patents, leaving the field free for others to improve transformer design.

In a more general context, under Ferranti the Grosvenor Gallery became a symbol of the British electrical industry's reawakening, ending four years of damaging inactivity during which little encouragement had been given to him or his contemporaries. Although Robert Hammond's Brush installation at Brighton had expanded steadily since 1881, by 1888 (as we saw in the last chapter) no local authorities or private companies had established a public electricity supply station. This illustrates the virtually unique nature of Ferranti's appointment, in stark contrast to the USA where Edison alone had installed generating equipment in 149 central stations by 1888. Hughes (1962) has argued that this British lag was largely caused by the lack of what he called 'the American spirit of go-aheadness', yet it is apparent from the study of efforts made by the early British electrical entrepreneurs that they were continually thwarted by an unresponsive environment. One could never argue that Crompton, Siemens, Brush, Hammond or Ferranti had not tried to overcome the fundamental problem of creating a larger market for electric lighting in Britain, but the existence of a cheaper standing competitor, gas, clearly presented a major obstacle to further expansion. This position was compounded by the legal restrictions imposed in 1882 and the unwillingness of financiers to support electrical activity, illustrating how the British scene had proved so obstructive that, even with such ambitious entrepreneurs, progress was seriously hampered. Only in 1886 did Ferranti appear to be turning the corner, after securing the Grosvenor Gallery appointment, and in time the others were to experience a similar increase in business as the economy levered itself out of the deep mid-1880s depression. This further emphasises the vital significance of Ferranti's twenty-second year to both his own development as an electrical engineer and the industry's slow rejuvenation. Even though his career had actually started in 1882, and the industry had come to the wider public's attention in that year, it was only from 1885 that Ferranti and British electrical engineering experienced any degree of continuity after three years of damaging stagnation.

During his first year as Chief Engineer to the Grosvenor Gallery station, Ferranti's energies and talents were pushed to their absolute limits, in the main because the directors were reluctant to replace all the original machinery with his own products. Modifications to the secondary generators were introduced to improve their performance but, as Sparks noted in one report, 'anyone having seen, heard or smelt the Gaulard & Gibbs transformers in action would never have thought of using such apparatus'. Juliana's diary for 1886 testifies to the arduous nature of the task, reporting how rarely Ferranti left the station before midnight in his attempt to maintain the supply of electricity to disenchanted customers when demand would have been at its peak. He was assisted in this work by a team of engineers recruited from among his family and friends. Sparks, of course, was his right-hand man, but also prominent was Ferranti's step-brother, Vincent Szczepanowski, who was eventually to supervise the installation of the Ferranti alternators in 1887, and later became Resident Engineer to the Grosvenor Gallery's successor at Deptford. Another relative employed was Ferranti's cousin, Hermann Donner. He acted as draughtsman, copying the sketches and translating them into working drawings from which Sparks and Szczepanowski could work. On the distribution side, a man with experience of cable-laying was hired, G.L. Addenbrook, while the former Chief Engineer to the Brighton station, H.W. Kolle, was recruited for his knowledge of practical operations. It was in the late 1880s that Ferranti first learnt the advantages of delegating responsibility for the more routine tasks of design and manufacture to subordinates, leaving him free to deal with the larger problems on any given project. In future decades he was to develop this idea of team research into one of the key features of his company's approach to the business of devising new products for the market, but it is important to note how in this period Ferranti was evolving his style after the more individualistic work of the early 1880s. What is also interesting about this multi-talented team is its youth: in 1886 Ferranti was twenty-two, Sparks twenty-one, Kolle twenty-two, Donner twenty-one, and Addenbrook twenty-six, while Vincent was by far the senior, at thirty-six. Perhaps it was their youthful exuberance which blinded them to the problems ahead, but it was this young team which was responsible for most of the pioneering work at the Grosvenor Gallery and its successor, Deptford, over the next seven years.

As Ferranti struggled to pull the Grosvenor Gallery through its early difficulties, and the load on the network increased, it became obvious that the station was not equipped for the challenges ahead. This prompted some serious debate among the directors, and it would appear that most were initially in favour of building a completely new station. In August 1886, Ferranti and Sir Coutts Lindsay even investigated various sites along the Thames, including the Stowage Wharf at Deptford, but after much discussion it was finally decided that the Chief Engineer should be asked to prove the viability of his system before larger sums were risked in an untried venture. Eventually, £20,000 was granted for the reorganisation of the Grosvenor Gallery station and, from late August, Ferranti started the process of scrapping the old equipment and replacing it with his own. This was no easy task because the engine room was located in the basement of the Gallery and the end wall could not be opened. It was also essential to maintain the supply to customers throughout the transitionary period in order to sustain the valuable goodwill built up by Ferranti and his team since the beginning of the year. As the journal *Industries* noted in November 1888: 'That a large building could be underpinned, foundations lowered to forty-six feet below street level, over 100 tons of machinery put in place, and some 40 tons of machinery removed ... without interruption in the regular supply of electricity reflects the highest credit on Mr Ferranti and his efficient engineering staff.' Further testimony to the difficulties of this conversion is that Ferranti did not take out any patents in 1886, one of only six years in his career in which this occurred. (The other barren years were 1899, 1914, 1916, 1917 and 1922, two of which were dominated by war work.)

Not only was the revised station a triumph of organisation and planning, it was also a highly lucrative contract for 'S. Z. de. Ferranti'. By the summer of 1887, they had supplied 150 transformers, each with their own switches, fuses and meters, and as new customers were added to the circuit more equipment was required. The most impressive feature of the new station was naturally the alternator Ferranti designed in 1886, because it was the largest generator built in the world to date. Ferranti had significantly modified the construction of his alternators by this time in order to increase their power, dispensing with the zig-zag armature in favour of independent coils bolted onto an iron core. This change is reflected in the capacity of the two Grosvenor Gallery alternators, the first of which was taking the full load by March 1887. Each had an unprecedented output of 2,400 V, stood nine feet six inches high, weighed thirty-three tons and, in relation to their size, were far more powerful than any of their contemporaries: a Gordon alternator of 1884 weighed twenty-two tons, but lit only 1,000 lamps, while the Ferranti machines were designed to light 10,000 lamps, but proved to have a capacity of 19,500 lamps. This says much both for the possibilities of central station supply by the late 1880s and for Ferranti's ability to design equipment for such a purpose.

The new Grosvenor Gallery station was actually completed in July 1887, and the team's success in overcoming the bad start of their predecessors in 1885 was confirmed by the growing business in electric lighting in the fashionable districts of the West End of London. When Ferranti had taken over as Chief Engineer, fewer than 300 lamps were connected on to the circuit, but by October 1887 five separate circuits had been laid, supplying electricity to 11,000 lights. The list of customers reads like an address book of the most fashionable clubs, theatres and residential districts around the Grosvenor Gallery, and even the Prince of Wales' residence, Marlborough House, was brought on to the network. This was the first time Royalty was known to have been linked with such an enterprise, giving the station an important position among the status-conscious classes of London. Covering an area from Regent's Park in the north to the Thames in the south, and from the Royal Courts of Justice in the east to Knightsbridge in the west, the Grosvenor Gallery network must have been a great source of pride to Ferranti and his team. It must have convinced Sir Coutts Lindsay & Co. that he had now established the viability of his high-tension AC system, and discussions on further expansion were already beginning by the time of the station's completion. Another factor to be considered was that although Sir Coutts Lindsay & Company were by far the largest supply operation in London, by 1887, as Parsons (1939) illustrates, new companies were beginning to appear on the scene. Given electricity's relatively high price, compared to gas, the west side of London was the most attractive market for electric lighting, and it was only natural that Sir Coutts Lindsay & Company's position in that area would be challenged forcefully in the near future. This made it imperative that extra plant be installed, but the Grosvenor Gallery could not be extended any further. Ferranti reminded the Board of their discussions in August 1886, when sites along the River Thames had been considered, and with this in mind Sir Coutts Lindsay & Company were voluntarily would up and their assets acquired by the newly-created London Electricity Supply Company on 26 August 1887. This new company, with the extremely large nominal capital of £1,000,000, was established specifically with the intention of constructing the first modern power station, at the Stowage Wharf in Deptford, a clear indication of the faith placed in Ferranti's skills by the Lindsays.

The Deptford scheme was to make the Grosvenor Gallery look paltry in comparison, not only because of the enormous financial investment, but also because Ferranti intended to establish the technological feasibility of the principles involved in high-tension AC on an even grander scale. Ironically, as he later admitted, he was taking his lead from the gas industry, electricity's arch-rival in the lighting market. Even since 1868, when the Gas Light & Coke Company had erected their large, centralised works at Beckton, the London gas industry had become a highly efficient system based on a rationalised structure of major plants feeding consumers through an integrated piping network. Ferranti hoped that the London Electricity Supply Company (LESCo) would provide the model for a similar development in the electricity supply industry. Certainly, there was great anticipation and incredulity surrounding Deptford, and as *Electrical Engineer* noted in October 1888: 'There is an air almost of necromancy, an Aladdin's lamp kind of effect upon the public and untechnical mind on learning the details.' Indeed, it was a scheme to make even the most optimistic sit up and take notice, for the initial plan was to install plant capable of lighting 250,000 lamps, with a view to expanding this capacity to 2,000,000 lamps when the demand arose, in the hope that LESCo would capture the largest share of the London market. The 'small' engine room — a misnomer when one compares it with the average engine room of that period, which would have been one-quarter its size — was to house two 1,500 hp Hick Hargreaves engines, each belt-connected to a 1,000 kW Ferranti alternator, while the 'large' engine room was to have four alternators of 3,000 kW powered by 10,000 hp Hick Hargreaves engines. This was an unprecedented scale for the time, not only from the perspective of Britain's unexacting standards in the mid-1880s, but also when one compares Deptford with the scale of activity abroad. In fact, Ferranti's machinery, and the entire system, was the largest in the world at that time, and only the construction of the Niagara power station in the 1890s, a scheme discussed in the next chapter, surpassed his engineering achievements, while in Britain it was another decade before a comparable scale was attempted. Another novel feature of the 10,000 hp units was the method Ferranti decided to employ in powering the alternators, attaching the armatures to the flywheel of the engine in an attempt to improve efficiency.

In the late 1890s, Ferranti supplied large numbers of such flywheel alternators to British customers, but at this time it was untried and contradicted much of the accepted practice in the country. The distribution system was also to be no less innovatory, increasing the 2,400 V of the Grosvenor Gallery to 10,000 V. Even Sir William Thomson was sceptical about this aspect of the scheme, but LESCo's engineer was not concerned with contemporary notions of standard practice, preferring instead to try new ideas in the hope that cheaper electricity could be supplied to the growing number of customers. Such an ambitious attitude was symbolic of the renewed optimism within the industry as a whole, but the scale of Ferranti's plans placed him naturally at the forefront of these developments.

There were, of course, many who shared his optimism, in particular the company's customers: the load had increased from 11,000 lamps in October 1887 to over 38,000 by the end of 1889. By September 1888, £535,000 of LESCo's capital had also been subscribed by twenty-eight shareholders, the most important of whom was Lord Wantage, Sir Coutts Lindsay's brother. Wantage had been invited on to the Board in 1888, after purchasing shares worth £220,000. This, as Parsons (1939) argues, reflected the courage of a man who had been awarded the Victoria Cross for bravery during the Crimean War. He was eventually to hold £550,000 worth of LESCo shares and carried the major financial burden during the company's darkest days. It was Lord Wantage who persuaded J. S. Forbes to join the company, bringing a businessman of wide experience in the electrical industry on to the Board. Forbes, besides being a successful railway manager, was also Chairman of the Edison & Swan United Electric Light Company and the National Telephone Company, indicating a firm belief in the future of electricity and, together with the Lindsay family, he steered Ferranti through the testing first four years of LESCo's history.

In retrospect, it seems strange that, just as the construction of the Deptford station was about to commence, Ferranti somehow found time to marry the girl he had been courting since the early 1880s, Gertrude Ince, and then leave the country for three weeks. Having spent most of the previous six years associated with Francis Ince, Ferranti had developed a close relationship with his senior partner's family, spending several summer holidays with them, during which he became closely attached to the youngest daughter, Gertrude. Their close affection for each other was to become a source of great strength to Ferranti in his business career and, after they were married in April 1888, Gertrude provided that domestic stability which was essential to his work. Ferranti could now afford to extend his financial commitments because, as Chief Engineer, LESCo paid him what was regarded as a high salary of £3,000 per annum. 'S. Z. de Ferranti' was also generating better profits by 1888, giving Ferranti a good income to support both his ageing parents and his own family. After the ceremony, the couple allowed themselves the luxury of a honeymoon, travelling for three weeks around France and Switzerland but, quite typically, Ferranti took the opportunity of visiting several electrical installations and, as we shall see in the next chapter, arranging certain business deals which were to serve 'S. Z. de Ferranti' well. Predictably, it was Deptford that dragged them back to London, and Gerturde (1932) remembered that this project was to dominate their lives at this time: 'The principal thing I remember during those first months of married life was Deptford, and again Deptford. We talked Deptford and dreamed Deptford.' For the next three years it was to become Ferranti's enduring passion as he waded confidently into the hazards of pioneering on the frontiers of technology, sweeping aside all warnings with the optimistic assertion that where Deptford would lead the way others would follow. Dubbed the Michelangelo of the scheme by the trade journals, because he alone was responsible for designing every aspect of the station with the sole exception of the roof, Ferranti hoped to establish a revolutionary new *modus operandi* for electricity supply stations.

The appearance of a growing number of competitors in London during the late 1880s naturally made it imperative that Deptford should be ready as soon as possible. Work had actually begun on the four-acre site in April 1888, using arc-lighters to facilitate progress on the night-shift, but even at this stage the team was hampered by poor weather which held up the construction of the building. Their first task was to lay a concrete raft, which acted both as a dock and as the station's foundations. A tramway was also laid along the full length of the raft, to convey the machinery and later the fuel. In line with contemporary practice, the pillars were cast on site, and slowly the building took shape alongside the Thames. By October 1888, twenty-four Babcock & Wilcox boilers had been installed, in sets of three, which exhausted through two chimneys, every two sets of boilers having their own flue. This well-organised boiler house was separated from the two engine rooms by a wall, and the foundations of these rooms were reinforced with iron girders and extra concrete to withstand the vibrations of the huge Ferranti machinery. The team worked energetically throughout 1888 but progress remained slow and, by the summer of 1889, when LESCo had already spent over £390,000 on land, buildings and machinery, it was becoming clear that the sanguine hopes of 1887 had been unrealistic. In his report of March 1889, J.S. Forbes could only hope that Deptford had 'turned the corner', but little was said about an immediate start-up.

While Ferranti and his team struggled with the early teething problems experienced at Deptford, they were also obliged to defend their ideas in a serious of public debates which culminated in a decisive judgement affecting LESCo's commercial viability. The wide interest aroused by the LESCo station had resurrected the 'Battle of the Systems', an argument which had raged in 1882 when both AC and DC camps had fought for the available business by decrying one another's technology. By the late 1880s, with the rejuvenation in electrical activity, electricity supply companies offering either AC or DC were engaging in a similar round of public statements. The lecture hall of the IEE resounded with the claims of both sides, the most famous event being the debate on R.E. Crompton's paper 'Central station lighting: transformers versus accumulators', which lasted for four sessions in 1888. Crompton was a fierce critic of the high-tension AC system, arguing that low-tension DC equipment was much more cost-effective. He admitted that the latter was more expensive to install, mainly because of the high cost of accumulators (batteries) and cable, but it was a proven fact that their running costs were lower, while the dynamos could be used to recharge the accumulators at off-peak times for use as a reserve. In the case of Deptford, Crompton felt that relying on a single set of generators was impractical, and alternators could still not be run in parallel, resulting in their having to be run for long periods with only a low load. Crompton later admitted in private conversation with Ferranti that over the long term AC would win out, but it was doubtful whether this system could compete with DC at that time. Appleyard (1939) has also quoted Ferranti's one-time associate, Sir William Thomson, arguing in 1888 that 'the DC system, with its great simplicity, and its thoroughly convenient and economical use of power, as well as light, is destined to predominate ...', illustrating the authoritative body of opinion against which Ferranti was fighting.

The DC camp's argument was greatly enhanced when T.A. Edison, the doyen of all electrical engineers, visited Deptford in 1889. This reflected the world-wide interest aroused by Ferranti's unprecedentedly large scheme, and not even a bad a cold could prevent the American entrepreneur from inspecting the station. When questioned about the pace of electrical progress in Britain, after seeing the Ferranti alternators, he replied: 'You may be slow to begin, but I must say that when you do go ahead you may even beat us.' However, as a firm advocate of the DC case, he was highly critical of Deptford's scale and refused to believe that 10,000 V was necessary or safe. Ferranti and Francis Ince replied to this criticism by publishing a pamphlet, *The Dangers of Electric Lighting*, insisting on the total safety of the high-tension AC system. They quoted the significant fact that at the Grosvenor Gallery there had been no fatal accidents using 2,400 V, while New York, where the voltage never exceeded 2,000 V, had the highest mortality rate from electric shocks. This trans-atlantic debate served to highlight further the work Ferranti was engaged upon. His task was to establish the viability of the AC system, technically and commercially, and refute the points made by men like Edison and Crompton, but when the Board of Trade decided to intervene in 1889 the debate took on a much more serious tone.

The Board of Trade's interest had been aroused by the resurgence of electrical activity after 1887, particularly in London, where several supply companies had applied for permission to light more than one borough. Although the 1882 Electric Lighting Act was repealed in 1888 and replaced by legislation more favourable to private enterprise – the lease was doubled to forty-two years and the Board of Trade was given greater discretionary powers to overrule obstructive local authorities – no permission had been granted to distribute electricity over two or more boroughs. The success of the LESCo project naturally depended upon access to as large a market as possible, in order to maximise the economies of scale possible using the Deptford plant, and the company had applied for permission to supply twenty-four different boroughs. It was also essential that their current position be defended, because other companies had actually applied for permission to compete in those areas already supplied by the Grosvenor Gallery station. The Board of Trade was most concerned about this state of confusion, and in April 1889 one of their experts on electrical matters, Major Marindin, was appointed to settle the matter of which companies would be given permission to light certain areas of London. In one respect, LESCo had already pulled off what many contemporaries saw as a remarkable coup, using Forbes's influence with the major railway companies operating in London to negotiate rights to lay cables along their lines. This obviated the need to seek permission from each local authority to lay cables under the streets, but the results of the Marindin Inquiry could still have grave implications for the company's future.

In view of the serious nature of the Marindin Inquiry, LESCo spared no expense in their legal representation, hiring Sir John Pember QC to plead their case. Pember was, in fact, first to speak and he gave a classic exposition of the AC case, hinging his main argument around an interesting fact: if a low-tension system was to light an area with 200,000 lamps, 408 tons of coal per day would have to be delivered by approximately 200 carts, adding further to the congestion in London's streets. Pember also produced the calculation that LESCo could save up to 13s per ton on the cost of coal delivered by their own colliers, compared to that delivered by the 200 carts. Marindin's report later conceded the point that the site of the Deptford station possessed 'the undoubted advantage over any others', but he was more concerned with the 'experimental nature' of the scheme which, in his opinion, militated against its locational advantages. One of the principal factors prompting him to draw such a conclusion was the explosion which occurred after his tour of inspection of the Deptford station, when a bend in one of the pipes supplying steam to the 1,500 hp engines burst. Not only did this accident result in the death of one man and the serious scalding of three others, it also materially affected Marindin's judgement on the future prospects for such projects. Of course, this was a failure in established steam power technology, but once again it was seen as evidence of Ferranti's over-ambitious tendencies. Ferranti later produced a design for steam pipe bends using a number of solid-drawn copper tubes, which the *Engineer* of March 1891 was to describe as 'beyond reproach so far as safely withstanding any working pressures' by then, but severe damage had already been done. Indeed, the 1889 accident, widely reported in the technical press, proved crucial in determining the outcome of the inquiry.

Although Pember had given a masterful description of Deptford's advantages over the other schemes on offer, as the inquiry progressed it became clear that LESCo were losing some of the more important debates. The explosion was undoubtedly a major influence on events, and it was also pointed out that the AC motor was generally inferior, making sure that the DC suppliers would have a foothold in all areas where industrial and domestice consumers might require power as well as light. Pember had stated that LESCo were not afraid of competing with DC systems, but, in most of the twenty-four areas for which it had applied, it was the AC system of the Metropolitan Electricity Supply Company that was the only competitor, and Marindin felt obliged to segregate them. In fact, as well as allowing DC systems to compete in their areas, Marindin only gave LESCo authority to supply electricity in thirteen areas, the majority of these being south of the Thames, an area not considered an attractive proposition at that time. Even more damagingly, he took away some of the customers already being supplied by the Grosvenor Gallery station, reducing the size of LESCo's potential and actual markets considerably. Although the Earl of Crawford tried to put a brave face on things by saying that LESCo was 'perfectly well satisfied with what he offered us', it is clear that Marindin had deprived the company of their main advantage of being a bulk supplier, not to mention a large number of their customers. The judgement made them simply one of a number of stations serving a limited area of doubtful potential in competition with DC stations using a well-established technology. This was obviously one of the major reasons why Deptford proved so uneconomic in the early years, but one could not deny the undoubted significance of the technological problems with which Ferranti was faced. In the period 1888–92 it was the 'experimental nature' of the scheme which prevented the full realisation of any commercial prospects LESCo may have possessed, and only in the 1890s did Marindin's judgement affect the company's performance.

By the end of 1889, LESCo had spent over £520,000 on land, buildings and equipment, but the station had still not come on-stream. They had even been obliged to purchase a £6,000 lathe to produce the larger alternators, because two of the major engineering companies in Europe, Krupps and Creusot, could only promise delivery of such unique components in three years. Such was the scale of this aspect of the scheme that the Deptford station was actually used to manufacture the 10,000 V armatures. The 10,000 hp engines had been successfully subcontracted to the Bolton firm of Hick Hargreaves, and the ingots used in their manufacture were said to be the largest cast in Scotland to date. This illustrates the unique nature of the project at this time, but the problems with the 10,000 hp units soon paled into insignificance when Ferranti turned his attention to the production of a suitable cable to carry the 10,000 V from Deptford to its four main substations. In the past he had always purchased cable from a specialist manufacturer — the India Ruber & Gutta Percha Company had supplied the 2,400 V cable for the Grosvenor Gallery network — but 10,000 V was beyond anybody's capabilities at this time and Ferranti was obliged to take on the design of the first viable high-tension cable himself. Sir William Thomson was highly sceptical about the prospects for the successful completion of such a project, enquiring 'Have you any practical trials yet up to such high potentials?' He agreed to provide Ferranti with some technical data on insulation materials, but it was to be in the field that the unprecedented problems of using such voltages had to be tackled.

In his 1885 master-patent for a system of high-tension AC distribution, Ferranti had incorporated the design of a tubular concentric main on which he was to base his Deptford cable. He had started the preliminary design work in July 1887, assisted by Sparks and Addenbrook, but it was the advice of Sir William Thomson which proved crucial in the adoption of a novel form of insulation. Sir William had advised Ferranti in 1886 that a firm of condenser manufacturers, Latimer, Clark & Muirhead, had experimented with the use of paper in place of the conventional dielectric, rubber, and after a series of tests by Ferranti it was noted that the addition of ozokerite (paraffin wax) would improve its insulation properties. Ferranti, Sparks and Addenbrook constructed a concentric cable to this design and discovered that paper was much cheaper and more effective than rubber. However, on laying the new main they found that this was not sufficient in itself to prevent induction problems. It was a major disappointment to the team, and by the end of 1889 the Ferranti cable was being replaced by one of Fowler-Waring make. The new cable was also concentric in design, but proved no more cabable of carrying the pressure and, as it was laid along railway lines, its outer coat of jute was susceptible to catching fire when hit by the sparks and cinders flying from the passing trains. As a result, tests on the Ferranti main were continued and, early in 1890, the team produced the simple refinement of an iron sheath in which the Ferranti concentric cable was to be placed, with an earthing system at the Deptford station, which proved capable of carrying the full load without induction problems. Thus was born the first viable high-tension cable, establishing Ferranti as a leading authority in the field and, as we shall see in the next chapter, earning him large royalties in the 1890s.

Owing to the lack of space in Ferranti's works, the new cable was actually manufactured in the Deptford station, using a laborious process imposed on them by the limitation that copper tubes could only be purchased in lengths of twenty feet. Four seven-mile lengths of cable were required to connect the station with its market, and such was the success of the training that, of the 7,000 joints made by the unskilled labourers hired to construct the main, only twenty-eight were found to be faulty. There were, of course, still many sceptics, particularly at the Board of Trade, and Ferranti was obliged to prove the safety of his cable by staging a most remarkable experiment. In the Deptford yard, one of Ferranti's assistants, H. W. Kolle, stood on the earthed copper plate holding an uninsulated chisel to the cable, while one of the foremen struck the chisel with a sledge-hammer, thus breaking the cable and activating the main fuse. Although Kolle later quipped that the experience had only been daunting because the foreman had never used a sledge-hammer before, his bravery and trust in Ferranti's design had convinced the officials present that the system should be allowed to go into commission. This marked a significant achievement for LESCo and Ferranti, and it was felt that success was now assured. Unfortunately, all this had taken until the summer of 1890, by which time LESCo had spent a total of £753,428. The cable had accounted for almost £150,000 of this, a disproportionate amount when one considers the expenditure on mains by the largest of the London supply companies, the Metropolitan Electricity Supply Company, of £106,000. (LESCo's distributing area covered approximately 3·4 square miles, while that of the Metropolitan Company covered 4·6 square miles.) Nevertheless, in his report of October 1890, Ferranti could claim that Deptford was 'working in a highly satisfactory manner', with the full 10,000 V on tap and flowing through his newly-laid cables.

Just at the time Ferranti and LESCo thought they had turned the corner and overcome all the major technical problems, an even greater setback occurred as a result of simple human error. On 1 November 1890, the Grosvenor Gallery station was finally closed down after five years' service which had seen even the liberally-proportioned Ferranti alternators hard-pressed to match the rising demand. The alternators were to be moved to Deptford to act as a reserve, and transformers put in their place, but while these changes were taking place the latter were housed in a wood-lined room above the boiler-house. It was during the move, early on the morning of 15 November, that a linesman made a disastrous mistake: bringing a fresh bank of transformers into service, he caused a 5,000 V arc at the plug switch and, instead of closing it, he withdrew the switch, causing the arc to expand and resulting in a fire that cost the company almost £20,000 worth of damage. Not only could the man have stopped the arc as soon as he had caused it, he could also have used the nearby circuit-breaker, or even have cut off the supply from Deptford, before the fire could take hold. Instead, he panicked and destroyed LESCo's key substation. Ferranti worked hard to replace the transformers, and by 26 November they had been installed in their correct position in the Grosvenor Gallery basement, but the haste with which they had been made was responsible for the failure of one on 3 December. When it burnt out, the load was transferred on to its fully-loaded neighbours, which also burnt, and the substation once again was in disarray. This resulted in a complete cessation of supply to the 38,272 lamps then on the LESCo circuits, and for over two months their customers were without electricity. It was a loss of goodwill the company could ill afford in such a competitive atmosphere.

In a limited sense, the fire was a blessing in disguise, in that LESCo was able to use the intervening period to put its house in order – to install the Grosvenor Gallery plant at Deptford, to complete the mains, and to rewind the 'small' alternators to generate at 10,000 V. The reason why they were forced to dispense with step-up transformers at the station and have the alternators generate at 10,000 V lay in the discovery of another practical problem unforeseen by Ferranti. In the course of laying the mains, he found that the voltage at the London end was greater than that at the Deptford end. In 1887, Oliver Heaviside, the prominent theoretician, had postulated that using both step-up and step-down transformers could induce this effect, and after the experience at Deptford this phenomenon became known as the 'Ferranti Effect'. Eventually, after the advice of J. A. Fleming had been sought, condensers and fuses were fitted into the circuit to compensate for this rise in voltage, but once again Ferranti was dealing with unique practical problems. Perhaps one might argue here that Ferranti's limited theoretical knowledge may have been a handicap, because he should have been aware of Heaviside's well-publicised experiments. This brings up the question of whether Ferranti was pushing the boundaries of electrical technology too far forward without even understanding the technical and commercial implications himself, a point to which we shall return later.

The two Grosvenor Gallery fires caused a crushing loss of business. When the supply was resumed in February 1891, only 4,800 lamps were left on the circuit and, although this had risen to almost 24,000 by May, only a small fraction of the capacity then available at Deptford was being used. In his report for 1891, Ferranti stated: 'I desire to call attention to the fact that from the commencement of your operations to the present time, no engineering or electrical difficulties whatever have arisen which I have not been able to overcome, and at present I know of no weak point in your system, and consider success to be now assured.' This confident, if over-optimistic, assertion, never to be misconstrued as boastfulness, was aimed at persuading the directors to continue on to the second stage of his scheme, the installation of the 10,000 hp units nearing completion in Bolton and Deptford. It was such naïve hopes which were to precipitate his downfall. At LESCo's fourth Ordinary General Meeting in March 1891, Forbes stated: 'We had, for good or evil, launched ourselves on a new path; we had no beaten path to follow.' For this reason, the shareholders had been willing to sanction the total expenditure of almost £750,000, but even contemporary enthusiasts like the *Electrician* were sure that severe economies had to follow. Indeed, it was decided very soon after that the commissioning of the 10,000 hp units should be postponed indefinitely, bringing into direct confrontation Ferranti's engineering aspirations and LESCo's commercial prospects. Allegedly, there were some fiery exchanges between Forbes and Ferranti − the former is reported as saying: 'You are a very clever man Mr Ferranti, but I'm thinking you're sadly lacking in pre-vision.' − but the businessman was to win this particular battle of wills, and the engineer decided to leave the company. At no point in the evidence is it suggested that Ferranti was sacked by LESCo, but leaving under such a cloud was obviously an unsatisfactory end to the relationship. Ferranti's successor as Chief Engineer, P. d'Alton, remarked in his first report that Deptford's designer had left 'by effluxion of time', and there can be no denying the commercial rationale behind Forbes's stance after such heavy expenditure on plant.

As soon as Ferranti had left LESCo, J. A. Fleming and J. Hopkinson were commissioned to report on the state of Deptford. Only two

such authoritative electrical engineers were able to follow the pioneering work involved in the project. They concluded that the system was
surviving 'on very thin ice' and, indeed, even after the completion
of the first stage, Deptford continued to have problems. Not all were
caused by the original nature of the system, with a fire under a railway
bridge burning through the mains in March 1892, but they confirmed
many suspicions that Deptford would never work. The worst incident
came in November 1891, when a long spell of fog and frost put the
station under such a severe strain that, as Forbes reported: 'The whole
thing came to a collapse: the dynamos, mains and everything went
wrong, and for four days or more we were without light.' Confidence
in Deptford was consequently never very high in the early 1890s and,
when combined with the market restrictions imposed by Marindin,
it is not surprising that the load was slow to increase. In fact, by 1894
LESCo was in such dire financial straits that Lord Wantage was
obliged to call in the Receiver, and not until 1905 did the company
pay its first dividend. By this time Lord Wantage had died, but LESCo
continued to operate along the lines of the 'Ferranti System', generating at 10,000 V, and, largely due to the talents of G. W. Partridge,
Deptford's Chief Engineer between 1892 and 1939, the station became
a classic example of the way forward in electricity supply. The alternators, transformers and cables installed by Ferranti remained in
service for many years to come, in the case of the latter for over forty
years. This supports the point made by Parsons (1939) that 'Ferranti's
instincts were right', and his engineering talent of great significance.
In fact, such was the growth experienced by LESCo in the mid-1890s
that Ferranti had been contracted to supply another 1,000 kW alternator in 1896. This still leaves open the question of Ferranti's, and
LESCo's, commercial acumen in starting the project at a time when
the market was only just emerging.

Had Ferranti's 'Bold Conception' been over-ambitious? In the case
of the first Grosvenor Gallery fire, was he guilty, as Hughes (1963)
argues, of 'overloading more than a switch; he was overloading an
entire engineering-human complex. He was asking for more from the
personnel than training and experience allowed.'? One might, of
course, ask Hughes how long it takes to train a man to operate a
switch, but clearly there is a case to answer here, because of the
financial burden imposed by the scheme and the poor service provided
by LESCo up till 1892. Hughes (1983) also criticises Ferranti for

'pursuing a course that briefly appeared to be the mainstream but took him out of the rising tide of technological change', in that Deptford used single-phase AC, while by 1891 this had been supplanted by the more efficient polyphase system developed in the USA by Tesla. Such an argument, however, employs excessive hindsight, because the polyphase system was only successfully demonstrated for the first time in the year Ferranti resigned from LESCo, while its practicality was still being debated up to a decade later. One should also note that the general principles of both systems were very similar, based on the need to build high-tension AC power stations where land was cheap and fuel and water readily available at a lower price. This laid the foundations for the development of electricity supply in the decades ahead as more people came to recognise the merits of such systems, and it is with this in mind that one might judge whether or not Ferranti was in the mainstream of technological progress.

Hughes, of course, is not alone in depreciating Ferranti's enterprise: Hannah (1979) argues that Deptford 'was well in advance of its time and LESCo paid the price of backing an unsuccessful project on the frontiers of technology', and Byatt (1979) suggests that 'Ferranti was moving in the right direction, but tried to go much too far too early.' These opinions fit into the conventional assessment of Ferranti's Deptford scheme; that it was a piece of intuitive dreaming with little commercial rationale in the context of the technological and legal problems at that time. As far as LESCo's shareholders and customers were concerned, Ferranti's desire to lead the way others were later to follow necessarily involved a long gestation period. One must not belittle the innovative nature of Ferranti's ideas, or ignore the successful solution of such unprecedented practical problems as the 'Ferranti Effect', but the fact that it took almost five years before a working system could be produced has reduced the credibility of the whole project in the short term. If one were to take a longer-term view, then Ferranti emerges with greater credit in establishing the viability of high-tension AC, but the loud complaints from customers call into question Ferranti's achievements as a commercial engineer.

One of the crucial points that arise from this debate is the restrictive environment in which Ferranti and other British electrical entrepreneurs were obliged to operate, in stark contrast to the USA with its expanding market and lack of tight legislative controls. Byatt (1979) is critical of the capacity planned for Deptford, quoting the fact that only by 1890 were there a million lamps in London, but he misses the point that Marindin severely restricted the size of LESCo's potential market, while the larger units were only going to be brought into service as the demand arose. This difficult market was an essential factor explaining the lag in the development of the British electrical industry in the 1880s, rather than any lack of enterprise on the part of such businesses as Crompton, Brush and Siemens. It is interesting to note that Hughes's article, aimed at disparaging the 'Bold Conception' of Deptford, comes but a year after he had criticised British electrical engineers for being unenterprising. Surely Hughes is guilty of condemning them for being unenterprising in one breath, and in the next condemning Ferranti for being too enterprising. As one of Ferranti's contemporaries wrote in the journal *Western Electric*, about an article in the *Electrical Review* criticising what it called 'The Dream of Ferranti': 'We hardly know what to think of the *Review*. Can it be that Mr. Ferranti incurred its displeasure by his bold and daring display of that characteristic so distinctly American?', that of enterprise. Nevertheless, as I have already intimated, although one can credit Ferranti with a significant contribution to electrical technology, such inspirational engineering failed to take into account the market and technological limitations of the time. The Deptford scheme was the symbol of the reawakening of the British electrical industry, in that after 1888 central stations became a feature of the British scene, and Ferranti deserves great credit for his foresight and technical skills, but as far as the consumers of London were concerned such a reputation was no substitute for a regular supply of electricity, while LESCo's shareholders waited in vain for a dividend for eighteen years.

But just how influential were Ferranti's achievements at Deptford? What impact did the construction of the world's first modern power station have upon the British electricity supply industry? The answer to these questions must be that Deptford, in the short term, had a limited impact on the type and scale of system employed. As Hannah (1979) describes, in the nineteenth century British electrical developments hardly amounted to a major leap forward of a Schumpeterian nature. Even after 1900 low-tension DC systems were still being installed, largely because the market for electricity dictated such unadventurous policies, while the powers vested in the local authorities by the legislation of the 1880s prevented further extension. In the first place, the municipalities had used their statutory powers to keep private enterprise out of the industry, but as more municipal electricity operations were started in the 1890s they worked to ensure that these important subsidising services for the rates would remain profitable by excluding competition. At the same time, the gas industry was still much more competitive because of the rationalised structure adopted since the 1870s and the production of more efficient gaslamps in the 1880s. Only the introduction of the first tungsten-filament light-bulb in 1911 brought electric lighting into line with gas. One of the few glorious exceptions to this tale of constraint was the enterprising work of Charles Merz with the Newcastle Electricity Supply Company (NESCo). This system, as we shall see in Chapter 5, had developed into the largest integrated supply system in Europe by 1914, using the same basic principles that Ferranti had been advocating at Deptford. In fact, Hannah (1979) has bestowed the title of 'English Edison' on Merz for his achievements with NESCo, at a time when the British electricity supply industry was 'a collection of huts and basements with clanking reciprocating steam engines supplying lamps within a relatively small radius'. Unfortunately, until the late 1890s the British electrical industry remained essentially small-scale because of this small market, and we shall see in the next chapter how damaging this was to prove when foreign competition intensified.

Ferranti, of course, was extremely disappointed at this poor rate of progress, arguing that the wider adoption of electricity would facilitate more rapid economic and social progress. Throughout his life he was an ardent advocate of the case for electricity, the plans for Deptford being the practical symbol of how he hoped the industry would develop. In 1894, in a speech to the Royal Scottish Society of Arts entitled 'Electrical Developments of the Future and their Effect upon Everyday Life', he first publicly explained the tenets of his 'All Electric' creed, emphasising the need to shape current projects to the requirements of the future. He was particularly concerned that more efficient use of energy sources should be pursued and, as we shall see in Chapter 4, pointed the way towards a greater use of turbines as a means of reducing generating costs. This would lead to more cost-effective generator design, but, above all, to much wider use of electricity in industry and in the home as the 'All Electric' ideal gained general acceptance. It was this ideal for which he worked during his career and, although up to the early 1890s British electrical engineers had received little encouragement from the market, after 1896 there was a steady increase in the use of electricity as local authorities and private companies invested heavily in electric lighting and traction plant. Hannah (1979) has shown that in 1895 only 38 GWh were sold in Britain, but by 1900 this had risen to 180 GWh, and in 1914 the figure stood at almost 2,000 GWh (GWh stands for gigawatt-hour and is equivalent to 1,000,000 kWh). This growth in consumption provided a much greater demand for generating equipment and accessories, stimulating the growth of electrical manufacturing in Britain after a decidedly hesitant start. We shall return to the story of the electricity supply industry in Chapter 5, but in Chapters 3 and 4 we shall examine how Ferranti fared as a manufacturer in what proved to be a vital phase in the history of British-owned electrical companies.

3

Lessons in management and finance

On the termination of his contract with LESCo, Ferranti was left to pick up the pieces of a career that had lost its rationale. The man dubbed the 'English Edison' by the technical press, and the unrivalled head of the AC camp in Britain, he was now obliged to revert to his original role as an electrical manufacturer after over six years as a supply engineer. Initially, he was to find this transition frustrating, having to compete on the open market against a large number of rivals for what was in the early 1890s increasingly scarce business. We shall also see how the sluggish development of the British electrical industry during the 1880s had put home manufacturers at a competitive disadvantage when compared with their American and German counterparts, in both technological and financial terms, and the slump of the early 1890s retarded progress even further. These problems were compounded by various structural weaknesses in the British market, in particular the use of consulting engineers by customers to design power stations, thus preventing the standardisation of production, and the predominance of municipalities in the market, bringing severe financial penalties into contractual negotiations. Clearly, British electrical engineers were not only faced with a serious threat from foreign competition in the 1890s, they were also hampered by enormous commercial and legal hurdles in the formulation of their business strategy, and in Ferranti's case one might add that the independent stand he always took regarding the ownership of the firm only exacerbated the situation. By the 1890s he had developed a fierce individualism which, although typical of British businessmen in general, was to prove financially imprudent in a period of illiquidity for the electrical industry. Ever since he had attained his majority in 1885, Ferranti had been able to canvass sufficient support from wealthier businessmen without allowing any dilution of his controlling interest in 'S. Z. de Ferranti', but in the 1890s his reluctance to raise much-needed equity capital damaged the prospects for more rapid growth. At the same time, he was anxious that foreign manufacturers should not dominate the British market and he worked vigorously to expand the range of products, to such an extent, indeed, that Ince complained of his desire to establish 'works that would be bigger than Palmers at Jarrow'. Ferranti also insisted on pursuing an expensive development programme to maintain the business's technological competitiveness. This may well have served to extend his reputation as one of the foremost British electrical engineers but, in the context of his individualistic business philosophy, it was a policy totally unsuited to the company's financial structure. This was the dilemma facing Ferranti in the period 1888–1903; whether to raise equity capital in order to improve his competitive position and lose control or whether to rely on self-generated funds and fixed-interest loans and see foreign competition come to dominate the home market. The failure to resolve this dilemma correctly ultimately resulted in his temporary exile from the business after 1903, yet it will become clear that market problems were to play at least as important a role in this outcome. The British electrical industry was to suffer badly in the changing environment of these years and, in reviewing Ferranti's business career, one must always balance any view of his company's performance with that of his rivals. Ferranti remained optimistic about the prospects for electricity, but such sentiments did not provide the basis for commercial success up to 1903.

Electrical developments were continuing apace all around the world by the late 1880s, and, undaunted by his experiences with the Deptford project, Ferranti also applied for the concessions to establish two of the largest electricity supply installations of the pre-war era, at Niagara and at Laufenburg. Either of these projects would have been more time-consuming than his contract with LESCo, had he been successful in winning one, and his efforts illustrate the desire to build more stations along the general principles of his London scheme. In fact, as early as 1887, surprisingly just as LESCo was being created, Ferranti was making plans to harness the water-power of the Niagara Falls. By 1888 he had even negotiated the concession to build a power station on the Canadian side of the Falls, but before the scheme went any further a combination of Deptford's start-up and the discovery that

British financiers would prove reluctant to support such a plan prevented progress. This was not the end of the Niagara Falls episode, however, because after the creation of the Cataract Construction Company in 1889 by a group of American engineers and financiers led by S. Evershed, a contest was held by the International Niagara Commission, awarding prizes of up to $22,000 to worthy tenders. Given his international reputation by this time, Ferranti was invited to submit a design, but it soon became apparent that Evershed and his partners simply intended to draft all the best ideas into an American-built station. This caused great indignation in electrical circles and, always a man to stand up and be counted, Ferranti complained in an IEE debate on the Niagara Falls project in 1893 that 'the designs were based on the unrecompensed work of the world'. Nevertheless, while his ideas languished on the drawing board, the Americans were able to make great strides in the construction of the largest power station in the world at that time, incorporating the new three-phase AC system. This is but one example of how foreign industries were gaining valuable ground on the British electrical industry, a theme which we shall now examine in some detail.

Undeterred by this lack of success in Canada, and by his rift with LESCo, characteristically Ferranti attempted to begin a new project in July 1891, applying for a concession to use the rapids of the Rhine at Laufenburg for generating electricity. Again, his major problem was the lack of financial resources, but he was able to interest a major German cable company, Felten & Guillaume – their connection with Ferranti will become clear later – and by 1892 a complete tender had been submitted. It is interesting to hypothesise whether or not Ferranti would have moved to Europe had he won the contract, but, after much bureaucratic wrangling and indecision, it was only in 1905 that the authorities made a decision. By that time, Ferranti had sold his interest in the scheme to Felten & Guillaume for £5,500, and his failure to win either of these two major contracts forced him to concentrate on the British market. Clearly, in the light of the effort he put into the Canadian and European projects, it would appear that Ferranti saw himself more as a supply engineer than as a manufacturer, but it was as the latter that he was to spend most of the remainder of his life. It says much for the state of the British industry that he was obliged to look abroad for suitable projects, because by 1891 the brief boom in electrical activity had subsided and manufacturers were once again forced to endure a deep depression. This would not have been so bad had it resulted in a consolidation within the industry, but we shall see that Ferranti and his contemporary electrical engineers were completely unprepared for the developments of the next decade.

Since its creation in 1885, the business of 'S. Z. de Ferranti' had increased considerably from the initial output of a few meters to a range of electrical equipment, including some of the largest alternators and transformers in the world. No figures for the entire period are available, but an unaudited balance sheet for 1889 claims sales of almost £55,000, illustrating the progress made since the first year of business. Indeed, the Grosvenor Gallery and Deptford contracts pushed the Hatton Garden works to its capacity and, by March 1888, the firm had moved to 46 Charterhouse Square, providing premises with five floors for the anticipated expansion in business. An interesting feature of the facilities installed in the new works was the Experimental Department, a vital aspect of all Ferranti's businesses in the future. In the past he had been obliged to carry out most of the development work on the shop-floor while a particular contract was in production, but after 1888 he was to make the Experimental Department the hub of the enterprise. It was from this area that most of the new products originated and, as we shall see in later chapters, he instilled this approach to development into the very fabric of his company, providing the ethos on which it prospered in the twentieth century. In an industry with a high rate of technological obsolescence, such an approach was essential if progress was to be made, and in the 1890s, with American and German competition beginning to bear down much harder, it became ever more important.

The Charterhouse Square premises were not only kept busy supplying LESCo with electrical equipment, 'S. Z. de Ferranti' was also able to export several alternators after agency rights had been negotiated in several countries in 1888. One of the business deals negotiated by Ferranti on his honeymoon had been the creation of Ferranti, Patin et Cie in Paris, in association with a French electrical enthusiast, O. Patin, and up to 1893 eighteen alternators were supplied through this agency to customers in France and Belgium. John Haskins in Boston, USA, and the River Plate Trust, Loan & Agency Company in South America were also commissioned to sell Ferranti equipment, but only the latter recorded any success after supplying three alternators to La Plata and Rosario. This was a policy of maximising the market opportunities created by the wide publicity surrounding the Deptford scheme, and the addition of business worth over £15,000 proved very welcome, particularly as each alternator was of a standard size, lighting 3,000 lamps. As we shall see later, this degree of standardisation was unusual, because most British customers insisted on the adoption of their own individual specifications, a habit which did much damage to the manufacturer's desire to produce at a profit. Another interesting feature of these alternators was that they were of the flywheel type, as with the projected 10,000 hp units at Deptford, and during the 1890s Ferranti based his reputation as a manufacturer of generating equipment on the supply of these units to many British customers. This illustrates the great value of these exports but, by the end of 1891, Ferranti and his team were devoting all their time to the business and the agencies were disbanded. Perhaps the agencies should have been maintained, because demand at home was falling off badly after 1891, providing Ferranti with major problems in the years ahead.

Ferranti's reputation in the British electrical market in 1891 was, of course, based almost entirely on his deeds at Deptford, and domestic purchasers were to prove hesitant in their attitude to his business in the next two years because of this station's unconventional nature. By 1892, of the fifty central stations in Britain and of the fifteen in the course of construction, Ferranti equipment was only being used at Deptford, Rochester and Glasgow, while, according to Byatt (1979), five firms (Siemens, Brush, Crompton, Mather & Platt and the Electric Construction Company) accounted for over three-quarters of the generator market. This left 'S. Z. de Ferranti' with a considerable mountain to climb in breaking into that market and, as the appendix reveals, output fell badly after 1891. The sharp downturn in demand for electrical products also coincided with the collapse of the engineering market and, after the standardisation of generator design, mechanical engineers were diversifying into the production of electrical machinery to utilise their spare capacity. Prices had already started to tumble in the wake of the trade depression of the early 1890s, but the entry of new producers into the electrical industry served to lower them even further because the usual practice, as the *Electrical Review* of July 1893 noted, was to tender at low prices until goodwill had been built up among customers. This intense competition was accentuated by the nature of the electrical market at this time, with contracts for the whole of the supply station equipment, from generators to switchgear, usually placed with one firm.

Byatt (1979) has shown that this single contract would occupy most of the manufacturing capacity of the typically small British electrical business but, as it progressed through the works, expensive spare capacity would be left idle, tempting firms to submit low tenders and thereby depress prices even further. In the gloomier circumstances of this period, such trends weakened the chances of British electrical manufacturers developing some of the new technologies, such as polyphase machinery and electric traction, which American and German companies were beginning to exploit at this time, and British investors were still reluctant to risk large quantities of capital in an industry with what appeared to be bleak prospects.

This combination of increased competition, scarce business and falling prices was naturally very damaging and, coming after the hesitant start of the 1880s, it extended the technological gap between the American and German industries on the one hand and the British manufacturers on the other. Passer (1954) has shown how, by the mid-1890s, the American electrical industry had experienced rapid growth, stimulating the creation of a duopolistic structure based around the interests of Edison's General Electric and the Westinghouse Company. Similarly, in Germany AEG and Siemens controlled their growing home market. In stark contrast, as we have already mentioned, the British electrical industry contained a large number of small businesses, a point well illustrated by Byatt's revelation (1979) that in 1898, when British manufacturers were reporting sales totalling no more than £100,000, General Electric produced profits of almost $400 million. Ferranti complained in a speech to the IEE in 1894 that there was nothing in Britain to rival such giants as General Electric and Westinghouse, and clearly the ruinous competition in a sluggish market would do nothing to enhance the possibilities of their creation. At the same time, the individualistic business philosophies of the era also held up greater concentration in the British industry, and Ferranti, as much as anybody, was reluctant to see control of his enterprise fall into the hands of others.

As a private partnership, 'S. Z. de Ferranti' could only rely on the capital resources accumulated from profits and salaries not claimed by the three partners. In 1890 this stood at £18,500, Ferranti having provided fifty-six per cent, Ince thirty-four per cent, and Sparks ten per cent. A £10,000 loan had also been raised from LESCo, but such paltry resources were woefully insufficient to cope with the large overheads borne as contractors to Deptford. The manufacturing plant at the generating station alone had cost over £17,000, while the fixed assets at Charterhouse Square were worth almost £21,000, and only generous prepayments by LESCo for the 3,000 kW alternator were keeping the business alive. This illiquidity worried Francis Ince deeply and, in a letter to Ferranti early in 1890, he argued that the only practical solution 'to mend the mess we are in' was the creation of a joint stock company with a major restructuring of the capital. Ferranti replied that this would 'take all [his] interest out of the work', a typical statement of the British attitude towards business management in this period, and instead he travelled around his business contacts 'practically speaking, begging' for capital to shore up his firm. He even asked Lord Wantage to join the partnership, but this entreaty, not surprisingly, fell on deaf ears and, in general, there were few people around who would invest in a company without either some say in its management or a guaranteed rate of return. Francis Ince was actually able to interest a group of financiers in the venture, but Ferranti was so fiercely opposed to their proposed role in supervising the finances that the deal soon fell through. This *impasse* was seriously jeopardising the future of the business and highlights the restrictive nature of British individualism in the context of business management. It was also causing personal problems for Ferranti and his family because, by the early 1890s, he was not only supporting his parents, he was also the father of two sons and a daughter. Gertrude (1932) remembers: 'There was little money coming in and it became essential to economise.' Ferranti was obliged to travel in third-class railway compartments on his begging errands, a sad reflection on his financial status at this time, while new clothes for the family were regarded as a rare treat.

It was only after much arm-twisting by Francis Ince that Ferranti finally agreed to the creation of a new company and, in June 1890, S.Z. de Ferranti Limited was registered with a nominal capital of £100,000, composed of 8,000 £10 ordinary shares and 2,000 £10 preference shares. This was a hollow victory, however, because Ferranti insisted on taking seventy-five per cent of the equity and over half of the preference shares, consequently preventing the separation of management and ownership common in public companies. They had simply created a 'private company', a form not legally recognised in Britain until 1907, but one which pervaded the industrial scene up to the First World War. It was, as Payne (1967) has argued of the private company form in general, 'a typical British compromise', illustrating once more the rift between the approaches of Ince and Ferranti to the business. The creation of S. Z. de Ferranti Limited highlights what was to become one of the vital factors influencing the company's performance over the next fifteen years, namely, the shortage of capital. No money was raised in 1890 and, after the break with LESCo in July 1891, illiquidity became even more acute as the order-book contracted and the £10,000 loan was recalled. Saul (1960) has argued that the small size of many British companies and the tradition of self-finance were serious hindrances to the growth of industry in general in the late nineteenth century and, as we shall see, this was certainly the case with S. Z. de Ferranti Limited in the 1890s. It was only the fortuitous creation of a cable company which temporarily provided the business with much-needed financial support after a rather traumatic phase in Ferranti's career.

After the termination of the LESCo contract in July 1891, the outlook for the company must have seemed bleak, with no guaranteed source of demand and a distinctly shaky financial base. It was just at this time that a Liverpool entrepreneur, J. B. Atherton, was trying to break into the well-established cable industry after creating the British Insulated Wire Company, of Prescot. Traditionally, with the exception of Siemens, the electrical engineering and cable industries had always remained separate, largely because of the latter's much earlier development during the telegraph boom of the 1840s and 1850s. Cable manufacture was a strong British industry with a technological ascendancy over its European and American rivals, facts which made many contemporaries sceptical of the prospects for success of Atherton's proposed use of American patents for paper-insulated cables. It became clear that these patents actually contravened those taken out by Ferranti for his high-tension main in 1890, but Atherton reacted quickly and by August an agreement between British Insulated Wire and Ferranti had been signed. In return for the use of his cable patents, past and future, Ferranti immediately received £6,210 in cash, as well as over £20,000 worth of shares in the company, amounting to almost one-quarter of the equity. He was also appointed on to the Board, while the company insured his life for £20,000 over the next five years to provide some compensation if their technical leader was to die. This was the beginning of a long and mutually-beneficial relationship which not only provided British Insulated Wire with the most important high-tension-cable patents in the world at that time, but also provided Ferranti with valuable finance just when it was required.

It was fortunate that Ferranti's technical genius in the design of a viable high-tension cable should provide the stop-gap for S. Z. de Ferranti Limited, in the form of the £6,210 payment. The British Insulated Wire agreement, indeed, proved crucial to the company's survival later in the decade, and in the early 1890s it brought Ferranti into contact with a wider group of business associates in the industrial north. It was to these people he was able to sell £20,000 worth of Preference Shares. Although it took him until 1893 to persuade all the investors to take the shares, over the intervening months money slowly flowed into the business. Among these investors were E. K. Muspratt, one of the leading figures in the chemical industry, who purchased £3,000 worth of shares, and Felten & Guillaume, the company Ferranti had been associated with on the Laufenburg project, who took £5,000 in shares. In the longer term, Ferranti's holding in British Insulated Wire's shares increased in value throughout the 1890s as the company established a reputation for the manufacture of high-tension cables. One of the most outstanding developments with which they were associated, using Ferranti's designs, was the introduction of flexible concentric cable in 1896, fifteen miles of which were supplied to LESCo. Within a decade British Insulated Wire came to be regarded as one of the strongest cable companies in Britain, and in 1900 an issue of £300,000 was oversubscribed. Ferranti's holding was worth almost £50,000 by 1899, acting as valuable security for the succession of overdrafts on which S. Z. de Ferranti Limited relied over this period. It is difficult to overestimate the value of such collateral, and certainly one can say that the company would have been in severe liquidity problems without this support. Ferranti had bolstered his finances in a very positive fashion by linking up with British Insulated Wire, but, on the other hand, this could not compensate for the lack of business in the Charterhouse Square works and he still had major problems to overcome in the market.

Although Ferranti was regarded as one of the leading figures in the British electrical industry, by 1891 S. Z. de Ferranti Limited could hardly claim to be a major supplier of generating equipment. As we noted earlier, five other firms accounted for over seventy-five per cent of the market, and for over two years Ferranti struggled in vain to secure a British contract to build a generator. In the meantime, the firm was finishing off its export orders won by Patin, but the mainstay of the business throughout this period was to be the Meter Department. It has already been established how vital Ferranti's meter designs had been in persuading Francis Ince to help create the new business in 1885, while the first contact with Sir Coutts Lindsay had been through the sale of thirty meters for use in the Grosvenor Gallery network. Over the next seventy years the Meter Department was to become the company's financial backbone, providing a healthy return on capital employed to subsidise the riskier ventures on which Ferranti and his successors embarked. In the 1890s, and in particular between July 1891 and the latter months of 1893, the company's main source of business was the production of meters, compensating for the lack of orders in other markets. It is interesting to note that, despite being the arch-proponent of the AC system, Ferranti was largely concerned with the sale of DC meters, an indication of the orientation of British electricity supply operations in this period. A Ferranti AC meter had been developed in 1888 and, while on his honeymoon, he had negotiated with two Swiss engineers, E. Paccaud and F. Borel, for the rights to use their AC meter manufacturing methods. In 1890, in association with A. Wright, the Chief Engineer to the Brighton station, Ferranti had also produced an AC meter particularly suited to measuring output from power stations, but most meters of this type were sold abroad. This was generally to be the case right up to 1914, the DC meter serving the home market and the AC meter customers overseas, and we shall see in the next chapter how an ambitious sales strategy was formulated to exploit this situation after 1905.

Since the introduction of Ferranti's mercury-motor meter in 1885, electrical engineers in the USA and Britain had paid much closer attention to this instrument, and by 1890 much more elaborate designs were available. Ferranti had adapted his meter in accordance with these refinements, maintaining his position as one of the leading British designers of reliable meters. In 1896, when the Board of Trade decided to insist on awarding its own seal of approval for electricity meters, the Ferranti product, along with one made by Chamberlain & Hookham, was the first to receive official acceptance. Chamberlain & Hookham were a Birmingham-based meter business which became Ferranti's main rival in this trade but, as we shall see in the next chapter, price-fixing arrangements were later negotiated between these two companies and the business remained profitable. Along with this insistence on continual design improvements, Ferranti also paid close attention to the organisation of meter production. Locating his business in Charterhouse Square, close by the Clerkenwell watch-making district, he had ready access to a highly-skilled workforce in the assembly of these intricate instruments. He even imported American machine tools for their mass-production, introducing a degree of standardisation and interchangeability unusual in British electrical machinery. To ensure that this investment would prove remunerative, a manager was appointed to supervise production, with the added incentive of a profit-sharing scheme. This brought an attention to departmental profitability which had hitherto been unnecessary in the company, a valuable managerial innovation reflecting the importance of the meter venture to Ferranti's business fortunes. As we shall see later, although Ferranti refused to dilute his ownership of the company, he was never slow to hire professional managers in his desire to delegate the more routine aspects of production, and the Meter Department was the first to benefit from this approach to the business.

Steadily, the meter business expanded after the move to Charterhouse Square, with output per week moving up from thirty in 1890 to seventy-two by 1892, and by 1893 the Meter Department occupied the top two floors of the works. Although there are few reliable figures available for this period, it is clear that Ferranti's technical ascendancy in meter design and production was successfully being converted into a highly profitable business: meter sales rose from approximately £16,000 in 1892 to over £25,000 by 1895 and profits averaged around £12,000 each year, rising to £15,000 in 1895. The appendix reveals just how crucial this proved to S. Z. de Ferranti Limited in this period, with meters accounting for a significant proportion of output and all the profits. In view of these statistics, it is clear that the meter market remained aloof from the general problems in the industry, a situation which can be explained by the growing number of customers connected to the generating capacity laid down in the years 1887–91. The Meter Department, above all, illustrates the value of having a bulk business to subsidise the riskier aspects of electrical engineering. In most cases, the major manufacturers used the light-bulb market in this way, and extensive price-fixing agreements were negotiated to ensure its profitability, but Ferranti concentrated on meters to provide the necessary funds for his other activities. It was this strategy which underpinned his company's growth over the next fifty years, giving the Meter Department a position of prime importance in the history of Ferranti's attempts to make a real impact on electrical technology. While Ferranti is perhaps best remembered for his much larger projects, it is vital to stress the role played by the relatively unspectacular meter in financing his work after 1891.

After the successful development of the meter, from the end of 1893 the heavier sections started to generate more orders. The real breakthrough came when Ferranti was able to convince Portsmouth Corporation that his equipment would prove viable. Originally, they had planned to install a low-tension DC system, and Professor Garnett of Tynemouth Polytechnic had been appointed to choose the best tender, but Ferranti pursued him around the country pushing the AC system. Again, Ferranti's persuasive nature proved successful and, in securing the contract to equip the Portsmouth station, he was able to embark on what was regarded as another radical departure from contemporary practice. Garnett was particularly attracted to the Ferranti scheme because of the proposed use of low-speed engines to power the alternator which was to be attached to the flywheel. This was the flywheel-alternator design employed in the proposed 10,000 hp units at Deptford and those exported to France and South America. Contemporaries were sceptical about this method because British central stations were usually fitted with low-capacity generators belt-connected, or directly-connected, to high-speed engines, but Ferranti was to dispel any criticism by producing the record low figure of $2 \cdot 3d$ per kW for the cost of electricity generation within the first six months of operations. This compared to the average cost in Britain in 1900 of $4d$ (Byatt, 1979). The flywheel alternators were of much more modest dimensions, compared to his Deptford generators, being of 212 kW capacity attached to cross-compound Yates & Thom engines, but their success set a new trend in British stations. It also provided the basic design for what came to be known as the Ferranti 'steam alternator', the trade name for his flywheel alternator, and over the next ten years much depended on its commercial success.

Portsmouth, rather than Deptford, can rightly be regarded as the starting point for Ferranti's period as a general manufacturer of central station equipment. Not only did it witness the steam alternator's first success in the British market, it also heralded the introduction of two new Ferranti products which were to feature prominently in the coming decade. One of the factors holding up the more extended use of AC in the nineteenth century was the continued use of arc-lamps for street lighting, and DC was over twice as effective for this purpose. Several engineers had attempted to overcome this drawback, but it was not until Ferranti introduced his rectifier at Portsmouth that a successful device was produced to smooth out the alternations in the AC flux to create an unidirectional current with the same qualities as DC for arc-lighting. It proved to be a major breakthrough, and by 1897 the *Electrician* was reporting that Ferranti rectifiers were rapidly replacing the Brush or Thomson-Houston arc-lighters formerly employed for street lighting. Of course, with the development of more powerful incandescent lamps, the life of the rectifier business was limited, but in the period up to 1903 Ferranti was able to sell large numbers of these machines, supplementing the regular generator and meter business.

A revealing fact about the Portsmouth station was the lack of any fire insurance cover. Considering the fire at the Grosvenor Gallery substation in 1890, this seems foolhardy, but such were the improvements in design and construction of Ferranti's switchgear that the *Electrician* of May 1894 claimed it was impossible by that time for his boards to cause similar accidents. His new system of containing the switch, cable receiver, fuse, ammeter and 'bus-bar of each generating unit in one cell of granite or marble has been described by H. W. Clothier (1933) as the basic principle on which high-voltage switchgear was to be based. Clothier had recently joined S. Z. de Ferranti Limited, and he was later to make a significant contribution to switchgear technology himself with the specialist manufacturers Reyrolle, having learnt his craft working with Ferranti in the 1890s. Ferranti also made another important contribution to high-voltage switchgear design when, in 1896, he first used oil to insulate the primary contacts, ensuring the total safety of this vital feature of a central station. This further highlights his contribution to the improvement of British power-station technology in this period, and by May 1900 the *Electrical Engineer* regarded Ferranti switchgear as essential equipment for all high-tension AC installations. Together with the steam-alternator, rectifier and meter, it provided S. Z. de Ferranti Limited with a viable product base for the immediate future. When the station was commissioned in June 1894, Ferranti was well-placed to exploit all the celebrations which usually accompanied such events at that time, and Portsmouth stands out as an important stepping stone in establishing his reputation as a central station contractor on more conventional lines than at Deptford.

The successes recorded at Portsmouth, indeed, came at an opportune time, because the electrical market was beginning to show signs of recovering from the lethargy of the early 1890s. As the appendix reveals, Ferranti's business was increasing at a much healthier rate after 1894, with the company making deep inroads into the former dominance of the generator market by the five major electrical companies (Siemens, Brush, Crompton, Mather & Platt and Electric Construction Company). The *Electrician*'s annual survey for 1896 showed that Ferranti alternators were being used in fifteen stations, compared to only three in 1892, and in the mid-1890s the company became one of the leading British power-station contractors. They even received a contract from the City of London Electric Light Company to build two of the largest generators in the world at that time, with a capacity of 1,500 kW. These were untypical of Ferranti's generator business in the 1890s, with the usual order being for between 200 and 600 kW machines, but they reflect his standing in the industry as a manufacturer of the most sophisticated equipment. Meter orders were also growing rapidly in the mid-1890s, pressing the Chaterhouse Square works to its full capacity as output almost reached the £100,000 mark for 1896. This forced the partners to consider the need for larger premises. Ferranti was obliged to shelve the plans to move early in 1896, because of an operation to have his appendix removed, but by the middle of the year the company was undergoing a major expansion.

The increase in business after 1894 was obviously the main reason why the company was considering a move in 1896, but another important consideration was Ferranti's decision to manufacture both the generator and the engine to improve the efficiency of the steam alternator. His fascination with steam engines had always been evident, ever since reading his step-brother's books in the early 1870s, but the growing popularity of the flywheel alternator forced him to think more seriously about the type he should use. We have already noted the convention in British power stations of using high-speed engines to run low-capacity generators, but as Ferranti's alternators were usually of a higher capacity he required a slow-speed engine, and few British manufacturers could provide products of the right quality. Characteristically, Ferranti was also convinced that he could improve on the efficiency of the steam alternator by designing his own prime mover, while dispensing with a subcontractor might maximise profit opportunities and eradicate all the delays and imperfections attendant on such a system. Although the profit projections were to prove woefully incorrect, as we shall see later, this was a diversification with great market potential, improving the company's competitive stance in the growing business in large generating sets.

Contemporaries agreed with this, placing several orders on its launch in 1894 and commenting favourably on the design of the engine, in particular the valve gear, at a time when foreign machinery was beginning to make a big impact on the British market.

With such a successful range of products on the market, and with the order-book building up nicely, Ferranti, Sparks and Ince decided to acquire new premises for the heavier departments. After an extensive tour of the Midlands and the North-west, eventually a four-acre plot in Hollinwood, near Oldham, was chosen. At first sight, the appearance and amenities of this district did not seem very promising. Hollinwood itself had a population of 9,000, housed largely in drab terraced rows, and dependent upon the local cotton mills and coal mine for its livelihood. Along the western edge of the common ran the Lancashire & Yorkshire Railway Company's line to Manchester, passing along its way two disused buildings formerly belonging to the Crown Iron Works, which Ferranti decided to rent. One of the major reasons influencing his choice was the availability of land for expansion, an essential factor in his planning, and in 1896 the long Ferranti association with this area had begun. It was an association which proved vital to the local economy in the decades ahead, with the Ferranti enterprise becoming one of the few reliable sources of employment when the cotton and coal industries began their alarming decline after 1920. Indeed, the move north was a wise choice for the company because land was cheaper, unskilled labourers' wages were thirty per cent lower than in London, and engineering skills were to be found in abundance for a concern moving into the production of mechanical, as well as electrical, products. These were important considerations for a cash-conscious company, and although the Meter Department remained in London for a further two years, duplicating some of the overheads, Ferranti's move to Hollinwood seemed eminently sensible.

The move into these larger premises, hereafter known as the Crown Works, provided Ferranti with the opportunity to organise production on a more integrated basis. Such was the lack of space in his previous facilities that the manufacture of the larger parts had usually been subcontracted out, but at Hollinwood a pattern shop, and later a foundry, was provided, along with adequate space for the assembly of the largest steam alternators. American light machine tools were also imported for use in the toolroom, where the special tools for the works were fabricated, making S.Z. de Ferranti Limited a self-contained organisation capable of supplying almost all its engineering needs. The whole works was powered by electric motors, providing greater flexibility in planning production routines, but it is important not to overstress the apparent efficiency with which the Crown Works was laid out. In fact, production was carried out with greater emphasis on the technical quality of the product, and works management would appear to have been almost non-existent. This, as Saul (1970) has shown, was typical of British engineering manufacturers, dominated as they were by customers who usually insisted upon products designed by consulting engineers, rather than accepting a standardised design offered by a company. It is also apparent, as Saul goes on to argue, that market factors alone cannot explain this aversion to organised production, because British engineers displayed an almost complete ignorance of the mass-production techniques developed in the USA by pioneers like F.W. Taylor. Taylor's 'Scientific Management' system, while extensively employed in various guises by American manufacturers, was generally regarded as inappropriate to the British scene by engineers who cared more for the quality of an engineering product than for the efficiency with which it could be manufactured. Ferranti regarded it as part of his duty to spend time on the shop-floor supervising the construction and testing of steam alternators, after having designed each machine specifically for an individual customer. Naturally, one is in danger of comparing 'best practice' in the USA with one medium-sized British manufacturer, but in this case, as we shall see later, despite the provision of expanded facilities, Ferranti failed to exploit the profit opportunities available with a larger business by improving the efficiency of production. On the other hand, free from the restrictions imposed by either the consulting engineer or the traditional emphasis on quality, American electrical engineers were able to standardise production to a much greater extent, giving their generating machinery a price advantage firms like S. Z. de Ferranti Limited found difficulty in matching. In the mid-1890s there was also growing friction between management and the unions in the British engineering trades, bringing further disruption to Ferranti's expansionary plans at a most unfortunate moment.

Ferranti's move into the Hollinwood area, in fact, did not begin very auspiciously, because in the engineers' lock-out of 1897 he took a resolute and uncompromising stand against the skilled members of his workforce. Within twelve months he had gained the reputation of being an employer who would brook no interference from workers in the management of his company. Ferranti had always been regarded as one of the better employers in London, taking a close interest in the welfare of his workers – Charterhouse Square employed approximately 400 men and women at its peak in 1896 – and each year he treated them to a 'Bean Feast' and Sports Day. On the other hand, his individualistic nature brought him into conflict with the power of the Amalgamated Society of Engineers (ASE) on the shop-floor, and he was only too willing to support his fellow-employers over such issues as the machine-manning question and the length of the working day. S. Z. de Ferranti Limited were actually members of the London branch of the Iron Trades Employers' Association, but after the move to Hollinwood, coincidentally just at the time the more powerful Engineering Trades Employers' Association (later known as the Engineering Employers' Federation and hereinafter referred to as the EEF) was created, the company joined this new, more militant alliance in the belief that a major conflict with the ASE was imminent. Wigham (1973) has described how, after a series of minor disputes earlier in the decade, by 1896 'every circumstance was leading towards the clash for forces', a situation which reached its natural conclusion in the following July with the lock-out. The two basic issues over which the employers and union fought were the eight-hour day and, more importantly, the fundamental issue of who should control the shop-floor, the skilled craftsmen or the management. With the introduction of more severe foreign competition in the late-nineteenth century, and the growing scale of production, employers were searching for more efficient methods of controlling production, diluting the influence formerly enjoyed by the artisan over such issues as the number of jobs retained by skilled men, and appointing supervisors to ensure their commands were obeyed. These changes were only introduced in a very piecemeal fashion and did not entail a complete turn-round in production methods, but they were sufficient to spark the major dispute of 1897. It was a dispute employers like Ferranti were anxious to win if they were to reduce what they saw as the ASE's damaging influence on company profitability.

As the dispute dragged on into the winter of 1897 and over 700 firms employing 48,000 workers became involved, Ferranti emerged as something of a local figurehead on the employers' side. The Oldham newspapers described what they saw as a scandal in his methods of inducing workers to join the company by paying their travelling expenses, hiring his own private police force to protect the men, and housing them in friendly parts of the town. In a speech to the Manchester District Engineering Trades Employers' Association, given in December 1897 and quoted extensively by Wigham (1973), Ferranti spoke proudly on the subject of 'How to Win a Lock-Out'. He was dedicated to the task of defeating what he described as 'socialism' because he felt such a movement 'would ruin the country'. He encouraged his colleagues to hire as many men as possible to keep their businesses going. The Crown Works was even staffed by students from the local technical colleges in the hope that Ferranti could 'show the ASE that they are not masters'. Inevitably, there was much dislocation in the company's output, but some progress was made, albeit at the expense of future notoriety. Francis Ince described Hollinwood as 'very warm', because of Ferranti's tactics, fearing violence and a direct attack on their property, but it was by pursuing such a course that the engineering employers succeeded in winning their fight. When the ASE came to a settlement in January 1898, they admitted defeat on the principal issue of managerial prerogatives, and the EEF was to reign supreme over the next few years. For Ferranti it meant a union-free works, with the ASE discouraged from recruiting there. This does not mean that he paid low wages — his rates were at least as high as those prevailing in unionised shops, and conditions were regarded as comparable — but his desire for complete independence in managing the company determined his attitude towards restrictive practices. In this he was again typical of most British businessmen and in the final chapter we will show how his labour relations strategy changed little throughout his life.

The 1897 lock-out was obviously a bad start to the company's move into larger premises, and the problems it posed added to the general strain as financial pressures once again made their impact on Ferranti's prospects for success. In the first place, he grossly underestimated the cost of equipping the Crown Works, calculating that only £20,000 worth of new machinery would be required. They actually spent £62,000 on supplementing productive capacity. Keeping two factories working, in London and Hollinwood, was also duplicating certain overheads, and in May 1898 the Meter Department moved north after Sparks calculated that over £3,700 per annum could be saved if Charterhouse Square was closed. This centralisation of manufacturing was a cost-reducing measure essential for a business almost entirely reliant on self-generated funds, and it is in this context that we must mention the vital importance of the relationship which S. Z. de Ferranti Limited forged with one of the most prominent North-west banks, Parr's. It has often been noted that British banks pursued a notoriously conservative policy with regard to industry but, as Cottrell (1980) has shown, Parr's were one of the great exceptions to this rule. After a vigorous phase of expansion and acquisition since the mid-1880s, Parr's had built up a large number of industrial customers and, when they had moved to Hollinwood, S. Z. de Ferranti Limited soon became one of them. In fact, without the overdraft facilities provided by Parr's after 1896, the company would have experienced extremely severe liquidity problems. Ferranti's British Insulated Wire shares were of paramount importance here, acting as security for this overdraft, but as it crept up to £34,000 in 1898 there were demands from within the company for a more radical solution to the cash-flow problems hampering faster growth.

As in the early years of the decade, the financial implications of remaining a private company were causing much friction between Ferranti and Ince, the latter reporting to Sparks in December 1897: 'I am most anxious at the way I am perpetually worried about this business'. Ferranti had actually suggested raising fresh capital in order to purchase the Crown Works, expand capacity and open a foundry; a strategy which typifies both his expansive attitude and his financial naïvety. Ince, of course, was astonished at these suggestions, and it was in this context that he complained of Ferranti's desire to 'build works that would be bigger than Palmers at Jarrow'. Ferranti's plans, with the exception of those for a foundry, were not implemented at this time, but by September 1898 it was finally decided to create 80,000 £1 preference shares and £100,000 in debentures. This was typical of the British electrical industry as a whole, because, as Byatt (1979) notes, fixed-interest securities dominated manufacturers' balance sheets. In fact, the Preference Shares were never actually issued, but by the end of 1899 £79,600 in debentures had been sold, placing a millstone around the company's neck at a time of acute illiquidity. (Interest on debenture loans must be paid each year, regardless of profitability.) As the appendix reveals, although no output figures are available, the company were recording larger surpluses through-out this period of buoyant demand for their products, but by relying so much on debentures and a bank overdraft they were storing up problems for the future. Illiquidity remained a major headache for the company, Ince reporting in January 1898 that one of his staff was 'either interviewing creditors or adopting means to keep out of their way nearly all day long'. This is firm evidence of the hand-to-mouth existence endured by S. Z. de Ferranti Limited in the financial sphere, bringing into question the policies pursued by Ferranti in his desire to retain control of the equity.

This need for working capital at least illustrates that business was expanding after 1894. Such was Ferranti's success in developing an attractive range of products that they had now entered the leading ranks of British electrical manufacturers, compared to their rather poor showing earlier in the decade. The electrical market had expanded so fast that, as Byatt (1979) describes, all the major companies were also expanding capacity, with total home sales rising from £1·3 million to over £4 million between 1897 and 1903. This was the kind of market stimulus for which the British manufacturers had been waiting since the early 1880s, but unfortunately this boom had two serious consequences for their long-term survival: firstly, the demand was largely from the municipalities, which were able to use their legal and market strengths to negotiate harsh bargains with suppliers and, secondly, the expanding market attracted a new generation of electrical manufacturers which completely swamped the older companies. The dominance of the local authorities was particularly damaging to company liquidity because the municipalities were able to insist upon the pernicious system of retaining payments for specified periods. This retention—money system, whereby the manufacturer would receive only small proportions of the contract price while installing the machinery, and up to twenty per cent was retained for as long as six months or until the equipment had proved itself, caused major problems. It naturally exacerbated the cash-flow problems of Ferranti's business, especially when the Treasury restricted local authority capital-spending programmes during and immediately after the Boer War, and delays were frequently experienced in receiving payments. The company's clerks were obliged to keep a meticulous record of payments due, and often the payment of creditors rested on the prompt arrival of customers' cheques. Given the financial problems related earlier, this naturally reduced the amount of cash available for product improvement, an important point to bear in mind when considering the company's financial performance, while recourse to Parr's Bank was often made for the payment of wages or for essential supplies.

Another feature of the local authorities' ordering pattern created problems for the manufacturer, in that the habit developed of appointing a consulting engineer to supervise the tendering for the contract, the detailed specifications to be fulfilled by the contractor, and the installation and running of the plant. This prevented the standardisation of designs, with each contract requiring fresh drawings, new templates and careful supervision of the manufacturing process to ensure that each machine came up to specification. It illustrates the high level of technological content which went into the manufacturing process in this period. In Ferranti's case, he was often to be found on the shop-floor supervising the actual construction and testing of the steam alternators with the small team he assembled in the 1890s to assist him in product development. The manufacturers attempted a combined attack on this problem by creating the Electric Plant Manufacturers' Association (EPMA) in 1898, marking the first attempt at collective action within the industry, but the local authorities were in such a strong position that little came of the negotiations with the powerful Municipal Electrical Association. In the USA and Germany, on the other hand, much greater co-operation between manufacturers and supply companies existed, with the former often being responsible for the latter's creation. This allowed the greater standardisation of designs and the use of lower skill levels in production to reduce the price of machinery. The *Electrical Review* of September 1900 complained that municipalities even presumed foreign-made equipment was superior to home-made plant because the Americans and Germans had greater experience in the manufacture of large machinery, and certainly there is plenty of evidence to support this view. As Byatt (1979) shows, by 1898 ninety-five per cent of the traction generators in the course of construction were American. Indeed, British electrical technology was well behind that of American or German companies, particularly in the areas of polyphase AC and electric traction, the two growth sectors of this period. Even Ferranti, although one of the loudest proponents of AC, was not yet producing polyphase equipment, largely because he could not raise the funds to develop a new range of products. By the turn of the century, this technical lag was beginning to have serious consequences for the survival of the British manufacturers as the foreign conglomerates started expanding into the home market.

If internal competition had been a major problem in the early 1890s, then in the latter half of the decade foreign competition was to accentuate the difficulties experienced by the British electrical industry. At first, imported equipment came to dominate the British market, in particular in the expanding electric traction business where American companies supplied most of the generators and motors up to 1901. This eventually stimulated the creation of subsidiary manufacturing operations by both the American giants, General Electric and Westinghouse, the former establishing British Thomson—Houston (BTH) in Rugby and the latter British-Westinghouse in Manchester. Among the established companies, Siemens invested in much larger facilities at Stafford, while Hugo Hirst's GEC built new premises at Witton, near Birmingham, in an attempt to move from their original base as a supplier of light electrical products into the heavier end of the market. It was also at this time that Dick Kerr & Company opened a new electrical factory in Preston, using American technology to exploit the traction and power station markets (Wilson, 1985). This activity completely dwarfed the older British-owned electrical manufacturers, with the three companies British—Westinghouse, BTH and Dick Kerr commanding total assets in the region of £5·8 million. (An equivalent figure for Brush Crompton, Electric Construction Company and Ferranti would only come to just over £1 million.) It brought a new crisis to an industry already heavily battle-scarred, revealing the full extent of the British manufacturers' shortcomings and putting in great danger their commercial survival.

The problems created by this new generation were obviously of major proportions, but to concentrate on these external pressures would only paint half the picture as far as S. Z. de Ferranti Limited were concerned. Just as important to this company was the self-created problem of Ferranti's refusal to raise equity capital and dilute his controlling interest. It was this attitude which starved the enterprise of much-needed finance and had already saddled it with the burden of fixed-interest securities. One might also argue that Ferranti's insistence upon quality, rather than quantity, production of steam alternators accentuated the liquidity problems by failing to meet delivery dates. Of course, the institution of the consulting engineer prevented greater standardisation, but Ferranti developed his products to such an extent that they were frequently late in installing machinery. The area where these strains were most felt was in the Engine Department. A total of £35,000 had been spent by Ferranti and his development team on this new design, and its mechanical construction was of a very high order, especially in the valve gear. As we have already seen, S. B. Saul (1970) has noted that British engineers were traditionally more interested in the technical quality of their product, as opposed to its commercial viability, and Ferranti was no exception to this rule, often spending so much time on the design and production of each engine that customers were able to impose harsh penalties for failure to deliver on time. Most steam alternator contracts included such penalty clauses, and in many cases they were enacted with dire consequences: in 1900 the Cardiff City Council was awarded compensation of £2,000 and the Ferranti steam alternator free of charge by the courts after the one-per-cent-per-week penalty had exceeded the contract price of £5,000. In fact, on the first twenty contracts for steam alternators, worth £60,000, the company lost £28,300, and even with a healthy Meter Department such a lame duck could not be sustained. Ferranti had produced a good design, but his manufacturing techniques had still not made much progress since his early dispute with Francis Ince in 1883, and in 1899 this prompted another major review of the company's position by a new figure of importance, the Chief Accountant, A. B. Anderson.

Although the company's major manufacturing site had moved to Hollinwood in 1896, Francis Ince's law business had kept him in London, and consequently the registered offices remained there. Ince only managed to keep in close contact with the works through the weekly reports of an Accounts Department set up for this purpose. This was a valuable addition to the controls introduced for the Meter Department in 1888, when a professional manager was hired to supervise production of their most profitable business, but in the late 1890s there were more important refinements in the company's managerial structure. It was clear that A. B. Anderson, as head of the Accounts Department from 1896, was slowly usurping Ince's former position as Ferranti's business adviser. Anderson had joined the company in 1894, at the age of twenty-four, and his ambitious style soon brought him into conflict with Francis Ince: he signed a company cheque for £200 in 1898, an act which Ince regarded as impertinent, and in 1899 Ince complained that he was no longer being sent all the weekly ways-and-means reports issued by Anderson, depriving him of the vital information on which executive decisions were based. It was also Anderson, rather than Ince, who was asked to carry out the review in 1899, reflecting his important role in company affairs by that time. In effect, after winning the battle of wills with Francis Ince, Anderson was exerting a considerable influence over financial matters, facilitating a major switch in the balance of power away from London and we shall see over the next two chapters how this benefited S. Z. de Ferranti Limited in managerial and strategic terms. One might actually guess that Ince was almost glad about this, because throughout the 1890s the financial worries had weighed heavily upon him. He had even threatened to resign several times, and finally in December 1899 Ferranti was informed that his senior partner was to retire from the business in March the following year. After eighteen years together they parted company the best of friends but, although Ferranti continued to consult his father-in-law on important matters, by 1900 it was clear that Ince's advice was no longer as crucial as it had been. Anderson and a team of clerks had replaced him and, given Ince's desire for orderliness in the pursuit of business goals, this was no doubt a source of great satisfaction to the man who had guided Ferranti through those early decades. It actually came as a great relief to Ince that he was no longer burdened with the task of supervising Ferranti's finances. He dearly loved his son-in-law, and was proud of the engineering achievements, but their conflict over business methods had been a constant source of trouble and the separation in 1900 saved a great friendship.

Although little evidence is available to support the point, it seems likely that Anderson's new status was responsible not only for Ince's retirement but also for the departure of Charles Sparks. In his report on the company, produced by the end of May, Anderson made the following diagnosis: 'There is no disguising the fact that for two-and-a-half years these Works have been entire chaos. There has been no system of works management, no method whatever of governing costs.' This was a direct attack on the current General Manager, Charles Sparks, and, as Anderson was also charged with the task of improving the organisation of the shop-floor after May, he was continually confronting the authority of Ferranti's junior partner. It may be a coincidence, but by December Sparks had left the company, to pursue a distinguished career as Chief Engineer to the County of London & Brush Provincial Electric Light Company in London, bringing to an end a relationship which had been responsible for some of the major electrical engineering feats of the previous fourteen years. Again, although Sparks' role as Ferranti's right-hand man had been vital up to 1896, after the move into the Crown Works his position had been reduced in importance by the appointment of managers for each of the major product groups, Engines, Alternators, Instruments and Switchgear, as well as Meters. Although all these men were engineers by training, this was the vital structural change of the 1890s because it extended the accounting innovations introduced into the Meter Department in 1888 to all the company's businesses. Each manager was obliged to submit weekly reports on departmental profitability, the value of work done, the amount invested in stocks, and the payments due from customers. A much more rational approach to cost accounting was also developed by Anderson, providing a financial structure which allowed rigorous central direction of budgets and overall company policy, if these managerial devices were employed effectively. Clearly, in view of the illiquidity of the enterprise and the poor works management, S. Z. de Ferranti Limited was not taking full advantage of the opportunities created by Anderson's reforms up to 1899, but there was to be a major overhaul of the entire organisation after May which brought solid financial rewards. Sparks clearly felt that this was an affront to his position in the company, and 1900 marks the beginning of a new phase during which the company underwent a drastic change in both personnel and management methods. From 1900, S. Z. de Ferranti Limited, although still owned by Ferranti, was then effectively controlled by professional managers, reflecting the more rational approach to works management and financial controls emerging within the company by this time.

It was obvious to Anderson that, apart from the poor works organisation, 'the want of money' and Ferranti's continuous development of the engine were the two main causes of the company's problems. He was particularly concerned that illiquidity had damaged their relationships with suppliers, to such an extent that discounts on bulk orders were rarely forthcoming. S. Z. de Ferranti Limited's method of paying creditors on a highly irregular basis had tarnished its credit-worthiness but, if fresh capital was injected into the enterprise, considerable savings could have been made in the purchase of essential raw materials. Ironically, because of the failure to co-ordinate supplies, stock levels had been kept at unrealistically high levels, compounding the cash-flow problem. Although profitability was improving after 1896, the high interest payments on the debenture loans and bank overdraft were sapping the life-blood from the company, and inadequate working capital was being provided for the expanding business. Of course, the retention-money system did not help the situation, but Ferranti's continued refusal to see his control affected by the introduction of professional investors held up greater progress. In July 1901 a new company was created, Ferranti Limited, to take over the assets of S. Z. de Ferranti Limited but, of the 100,000 controlling £1 Ordinary Shares issued, ninety-six per cent were alloted to Ferranti, and only 60,000 £1 Preference Shares were actually sold on the open market. (The nominal capital was set at £400,000, composed in equal parts of Preference and Ordinary Shares, but with only the latter possessing the right to vote at company meetings.) Ferranti Limited was simply another private company, raising very little capital − the bulk of the £60,000 raised in Preference Shares was used to repay the bank overdraft and finance the restructuring − and strengthening Ferranti's grip over the business. In fact, 120,000 Preference Shares had actually been issued in 1901 but, in the wake of the Boer War, investors were reluctant to venture much capital in a company so tightly controlled. This further highlights the problematic relationship with the financial sector and, as a result, recourse to overdrafts and debentures was once again forced on Ferranti. Anderson had been able to transfer most of the profits made since 1898 into a General Reserve Account, putting aside valuable finance for the troubles ahead, but this was only capable of keeping the company afloat for another two years.

Another reason why the new company failed to impress potential investors was the Board's composition. After the departure of Sparks and Ince, Ferranti had decided to elevate Anderson and three of the departmental managers to the position of Director, but this gave the impression to outsiders that engineers dictated company policy. It was an impression which probably came very close to the truth, and Anderson persuaded Ferranti to appoint some prominent local businessmen on to the Board in order to improve the company's visibility. In this context, Scott Lings was invited to join the company, bringing his wide reputation in cotton promotion − he had largely been responsible for the creation of Fine Cotton Spinners and Doublers Association Limited in 1898, one of the largest companies in Britain at that time − to enhance Ferranti's position in Manchester financial circles. Lings had recently published a book on company promotion, *Trade Comination*, in which he argued that owner-managers should hold at least one-third of the equity in their enterprise to convince potential investors that it was a going concern. This concurred with Ferranti's business philosophy, giving them some common ground on which to build a working relationship. Certainly, Lings' presence on the Board seemed to inspire greater confidence in Ferranti Limited because, in 1902, a further 60,000 Preference Shares were sold and later Lings placed a large package of Ferranti's Ordinary Shares. At the same time, the departmental managers had been asked to relinquish their position on the Board and concentrate on controlling their businesses on the lines set down by Anderson. The central control of management had also by now been passed to Anderson, who was appointed Joint Managing Director with Ferranti in 1902. It was a managerial structure which presented a better balance between engineering and commercial interests, but we shall see later how this was all brought to nothing by the market pressures exerted by the new generation.

While Anderson had been able to effect some significant changes in the management of the company, he was also successful in improving the efficiency of engine production. This department had been the major obstacle to greater profitability, in particular because Ferranti had, according to Anderson's report, 'put an experimental engine on the market'. We have already noted the heavy losses made on engine production up to 1899 but, after Anderson had introduced more effective cost controls and forced Ferranti to allow production to continue unhindered, a remarkable turn-round was experienced. The department recorded its first trading profit in 1900, albeit of only £3,521 on sales worth £36,500, and contracts were going through the factory in an average of almost nine months, compared to over twenty-three months in 1898. The *Electrical Engineer* of May 1900 noted that, in a review of the Crown Works, the main erecting shops showed 'admirable organisation', with strict attention paid to efficient production whenever possible. The nature of the British electrical market naturally prevented standardisation and mass-production, but techniques had been significantly overhauled after 1899, instilling greater confidence in the company among customers. Orders for four 2,500 hp steam alternators from the London County Council, two 1,100 hp units for the Capetown Corporation in South Africa, and eleven other contracts with British customers between 1899 and 1901 substantiate this point. In fact, by 1900 Ferranti generating equipment was used in twenty-eight of the seventy-nine supply stations in Great Britain, illustrating their continued progress since the problems of the early 1890s. The company seemed to have overcome the market's initial hesitancy and established a leading position in the central station business, but the financial pressures were to continue after the creation of Ferranti Limited, and the new generation of electrical manufacturers was beginning to make its presence felt in the worst possible manner.

When the company was reconstructed in 1901, the *Electrician* of July 1901 remarked that the firm 'gave every appearance of a sound and promising investment', with work-in-hand having risen from £99,000 in 1899 to just over £120,000. Output figures are not available for this period, but it is clear that meter production was still highly profitable, with the number sold reaching almost 7,000 by 1901. Such was the success in selling Ferranti switchgear that new premises in Hollinwood, the Windsor Mill, were rented and £35,000 was spent on new plant for the production of what was regarded as the leading design on the market. The workforce had also expanded considerably since 1896, with over 1,000 men and women working in the Crown Works by 1901, compared to 400 at Charterhouse Square before the move. In 1902 even the reissue of 60,000 Preference Shares went through successfully, providing the capital to finance the expansion of business, but it would unfortunately be misleading to over-emphasise these solid indicators of progress. It is much more realistic to say that, despite the increase in business, despite the reorganisa-tion of engine production, despite Anderson's managerial innovations and despite record profits in 1901 of over £25,000, Ferranti Limited was to experience even more serious cash-flow problems than S. Z. de Ferranti Limited had done in the early 1890s. Just to indicate the pressing nature of the company's problems by 1902, it is interesting to note that £25,000 of the money raised from the sale of Preference Shares in that year was used to clear the accumulated overdraft pro-vided by Parr's Bank. The balance sheets also record that, despite the accumulation of over £50,000 in the General Reserve Account by 1899, this had fallen to only £6,000 by 1902, such were the demands on finance in the early years of the twentieth century. Once again, another financial crisis was looming to hold up Ferranti's progress as an electrical manufacturer.

The principal determinant of the company's problems after 1901 was the unremunerative nature of the steam-alternator business. Although improvements had been introduced into the organisation of the Engine and Alternator Departments and orders were still being placed, prices dropped alarmingly as the new generation of electrical manufacturers attracted custom by offering more advantageous terms. In the case of Dick Kerr, their policy was to break into the generator market at any cost, the Works Committee deciding in 1902 to tender 'on all inquiries at a sufficiently low price to warrant our quotations being considered' (Wilson, 1985). There could be no hope of forestalling the entry of these companies because as we noted earlier, they were so much larger than the older British manufacturers. The *Electrical Review* of December 1900 had warned that prices would begin to tumble once the new factories reached full production but, to make matters worse, there was a sharp recession in Germany after 1900 and there is evidence in the tender columns of the trade press that companies like AEG deliberately undercut British prices to off-set their problems at home. Anderson calculated that, whereas in 1899 S. Z. de Ferranti Limited had been receiving on average £7 per kW for its steam alternators, by 1901 this had plummeted to £4 10s per kW. The retention-money system also continued to cause problems, Ferranti reporting to his wife: 'It seems nothing but delay in collecting what is owing to us.' More importantly, the fall in prices was accompanied by an even more disastrous decline in orders placed with Ferranti Limited after 1902, such was the success of the new generation in controlling the generator market. Byatt (1979) shows that, while British-Westinghouse, BTH and Dick Kerr were able to increase their share of this market from six per cent between 1893 and 1898 to fifty-two per cent by the period 1904–8, Ferranti Limited's share fell from fourteen per cent to only one per cent, respectively. The sheer force of competition was imposing a severe strain on the company's chances of recouping any return from its investment in this business. At the same time, profits were falling drastically, leading Ferranti into the crisis measure of selling 39,000 Ordinary Shares in 1902. This was a development which says a good deal about the situation, but in many ways it was simply the last gesture of a man overwhelmed by the changing circumstances in the market for his largest product.

In analysing the reasons behind the growing crisis in Ferranti's affairs, one is faced with a combination of factors, both internal and external, which seemed to have played equally important roles. Ferranti's desire to expand and diversify his business, while retaining the bulk of the equity, was clearly a major constraint, depriving the company of working capital and forcing a reliance on fixed-interest loans which drained the profits. On the other hand, the sluggish growth of the British electrical market up to 1896 had prevented the accumulation of reserves and, by the time demand was reaching more remunerative levels, foreign, or foreign-owned, companies were so much more competitive that British-owned enterprises were unable to capitalise on the opportunities. Indeed, Ferranti Limited's experiences were by no means unique among the older

British manufacturers, with most companies suffering a similar decline in business. They too had expanded their capacity in the late 1890s — Brush moved into new works at Loughborough and the Electric Construction Company invested heavily in a factory to provide similar units to Ferranti's steam alternator — but, as the *Electrician* reported in October 1904, the intense competition from the new generation had eroded their market position to such an extent that all were leaving the heavy electrical business to concentrate on specialised areas. One might also point out that British-Westinghouse, GEC and Dick Kerr experienced considerable financial difficulties up to the First World War, largely because of the market's failure to maintain its expansion after 1903, with severe price competition leading to heavy losses. Nevertheless, it is clear that Ferranti accentuated these market problems by refusing to allow any divorce between his ownership and control of the company, and the decision to sell Ordinary Shares in 1902 came too late. By that time, Ferranti Limited was so heavily burdened with debt and debentures, and the reserves were at such a low level, that it was only a matter of time before the company would succumb to the market pressures. In this context, it seems highly surprising that Ferranti was actually able to sell 39,000 of his Ordinary Shares in 1902, but this is evidence of Lings' assistance in this area. The bulk of the shares were, in fact, purchased by several of Lings' associates in the cotton industry, including W.H. Houldsworth, A.H. Dixon and L. Rivett, providing an infusion of capital at a crucial time. In view of the cotton industry's problems at this time, largely stemming from overcapacity and underinvestment in new plant, it also seems surprising that such businessmen were willing to invest so freely in an ailing electrical enterprise. It is, however, clear that this is not the only example of such diversification of interests within the cotton industry, because Dick Kerr relied heavily on capital provided by the Manchester cotton manufacturers S. & J. Prestwich, and many Lancashire tramway operations benefited from similar sources (Wilson, 1985). Such an investment strategy was clearly being prompted by the very problems facing cotton at this time, the businessmen concerned realising that more secure returns might be earned from an expanding industry than from one undergoing major changes. Unfortunately, though, the investments in Ferranti Limited were nowhere near as successful as those in Dick Kerr, and one can only guess how much damage this would have done to Lings' reputation.

The plunge into the red during 1903 was beginning to stretch the patience of Parr's Bank as the overdraft increased to almost £38,000 and it seemed inevitable that the debenture interest would have to be passed. Hector Christie, the Manager of Parr's Bank in Manchester, was one of the trustees for the debenture-holders and, acting on their behalf, in September he decided to call in the receiver to review the company's position. An approximate balance sheet was drawn up for the year to September 1903 and it was discovered that a loss of £58,557 had already been made, dictating a series of major changes which altered the whole structure of the business. It is to the credit of Ferranti's attempt to produce a competitive range of electrical machinery and instruments that at no time did the liquidators seriously consider dissolving the business permanently, and almost immediately two receiver-managers were appointed to take over the management of the Crown Works from Ferranti. At a meeting of the company's creditors in November it was agreed that their interests would best be served if the more profitable meter, switchgear and instrument business could be retained and, on this streamlined basis, the company was to become commercially viable after 1905. Nevertheless, after the receiver-managers took control in September, Ferranti's position at the head of the business was increasingly questioned, bringing to an end his period of unchallenged hegemony as Chief Executive of Ferranti Limited.

In looking around for two appropriate men to act as receiver-managers, Hector Christie was convinced that only trained chartered accountants would be capable of relieving the financial chaos in Ferranti Limited's balance sheet. Closely connected with Parr's Bank in Manchester was Arthur Whittaker, a practising accountant who was already conversant with the company's affairs when he was appointed as one of its receiver-managers in 1903. Over the next fifty years, Whittaker was to remain a director of Ferranti Limited, acting as senior adviser on financial matters, but his fellow-receiver-manager in 1903, A.W. Tait, was to play an even more important role in the company's immediate recovery. Although still in his twenties, Tait was a partner in the prominent London firm of accountants, G.A. Touche & Co., and, as Davenport-Hines (1986) has noted, he was already gaining a reputation in the City as 'a force in industrial finance' by the time of his appointment as receiver-manager to Ferranti Limited. Besides his generally-acknowledged organisational skills and business acumen, Tait was very concerned about what he saw as the 'curse' of individualism which he felt had dogged the progress of British business over the last generation. This naturally brought him into conflict with Ferranti's business philosophy, but such was Tait's power as receiver-manager that the founder of the company found it increasingly difficult to influence policy-making decisions after 1903. On the other hand, in the first report submitted by Tait and Whittaker at the end of September, they did accept that 'the management of the various departments is in the hands of a highly trained staff'. They also noted how the problems of engine production had been successfully overcome since 1899, confirming the points made earlier about the reorganisation of the business, but their dislike of Ferranti's financial policies inevitably laid the basis for vitriolic arguments. As professional accountants, Tait and Whittaker were concerned with liquidity, not personal control, and their view of Ferranti Limited's future was distinctly different from Ferranti's.

By the early months of 1904 it had become patently obvious that Ferranti, Tait and Whittaker would never be able to work in harmony, in particular because the receiver-managers planned to rationalise the product range. Their first report had already drawn attention to the vital role played by the steam alternator in precipitating the company's collapse and, in a second report, they recommended the termination of this business. It was clear that production of such large machines was impractical, given Ferranti Limited's financial resources, while the competition was simply overwhelming. This brought to an end Ferranti's long association with the design and manufacture of generating equipment which had seen the production of some of the largest machines in the world. Ironically, the last steam alternators were delivered one month early, to the Capetown Corporation in 1904, but by then prices were so unremunerative that the company could no longer sustain the heavy losses. Tait and Whittaker had also decided at a Board meeting in December 1903 that 'it was considered very desirable in future not to take up new branches of business', a direct attack on Ferranti's pursuit of expansion and diversification since 1891. More specifically, this was intended to curb expenditure on two high-risk projects Ferranti had started in 1901, the design of turbines and electrically-powered textile machinery. More will be said about this work in the next chapter, but by the end of 1903 almost £5,000 had been spent on experiments in the Crown Works and a similar sum on taking out patents covering the turbine in Europe and North America. Apart from distracting Ferranti and his team, this was cash the company could ill-afford at the time and it is unlikely that sufficient capital could have been found to produce these bulky machines. In one sense, given the growing popularity of the turbine and the many contemporary predictions that it would soon replace the reciprocating steam engine, it was quite sensible for Ferranti to venture into this field in order to provide a replacement for the steam alternator. No such logic, however, was attached to the textile machinery project, while in the eyes of Tait and Whittaker Ferranti had overstretched the business, and their strategies were clear challenges to his approach. Of course, in law Ferranti was no longer in control, disdainfully accepting the inevitable changes at the Crown Works introduced by the receiver-managers. After over twenty years building his company into one of the most respected electrical manufacturers in Britain, Ferranti was being deprived of its management.

The company was finally reconstructed in February 1905 and, although the name Ferranti Limited was retained, it was completely different in structure from that registered in 1901. Ferranti retained a seat on the Board, but he was only to act in an advisory capacity on technical matters, while Tait became Chairman. The company's nominal capital was also drastically reduced from £400,000 to £130,000, made up of 60,000 six per cent Cumulative Preference Shares of £1 (allotted to the creditors to the nominal value of their debts), 60,000 £1 Ordinary Shares (allotted to the former preference shareholders) and 10,000 £1 Deferred Shares (allotted to Feranti in lieu of his ordinary shareholding). As each share carried a vote, Ferranti was now reduced to a relatively insignificant position within the company and, in fact, Tait ensured that he was virtually excluded from the Crown Works after 1905 because of his alleged obstructive influence on profitability. To represent the interests of the creditors, who now held all the Preference Shares, J. M. Henderson was elected on to the Board in 1905. He was a Director of Thomas Bolton & Sons, the copper smelters, and as a chartered accountant and barrister he was to play an important role in the next decade. In this context, Ferranti Limited stands out as an excellent example of the growing importance of professionally-trained businessmen, and especially accountants, in the management of manufacturing companies in the two decades preceding the First World War. A. W. Tait was one of the more prominent members of this emerging group which was beginning to contribute significantly to the improvement in British business practices at a time when foreign competition was highlighting the antiquated individualism which had formerly dominated. At Ferranti Limited, although the founder had introduced professional managers to supervise production, and a system of departmental reporting had been developed to improve accountability and the flow of information, after 1903 much more rigorous controls were introduced as a means of tightening up the company's performance. The financial embarrassment of 1903 and the depression in electrical activity prevented more extended use of outside sources of funding, but Tait and Whittaker were to use professional investors and internal resources more imaginatively than Ferranti would have done, leading the company out of its demise and into a rosier future. It was a future, though, in which Ferranti was to play only an indirect part up to 1914.

The story of Ferranti's business ventures in the period 1888–1903 illustrates quite graphically the nature of the problems experienced by British entrepreneurs in establishing such new industries. We have seen in the two earlier chapters how the environment proved extremely hostile in the 1880s and, while the demand for electrical equipment increased at a faster rate after 1896, the damage had already been done in failing to provide British manufacturers with adequate financial and technological resources for the struggle to compete with stronger foreign conglomerates. Of course, in Ferranti's case one must not over-emphasise the role played by external factors because his desire for complete independence of action undoubtedly contributed to his problems. This philosophy was typical of that espoused by many prominent British businessmen, and he was by no means exceptional in holding such views, but the electrical industry demanded a more open approach to finance because of its rapid rate of technological development and the relative backwardness of British companies. On the other hand, however, it is not absolutely clear whether or not the City would have provided the necessary finance to improve Ferranti Limited's position, given the electrical industry's poor relationship with investors since the 1880s. Looked at in perspective, Ferranti seemed to be developing a competitive strategy for his business which would equip it with a product range capable of competing with both British and foreign electrical manufacturers. That he wanted 'works that would be bigger than Palmers at Jarrow', as Ince noted, was part of his general concern, shared in many parts of British industry, that foreign manufacturers were beginning to dominate the world's markets in such growth sectors as electrical engineering. Neither was he slow to hire professional managers to supervise both production and such functions as accounting, introducing departmental reporting as a key feature of the company's structure after 1896. Nevertheless, his fierce individualism and continual desire to develop new products for their technological value, rather than their commercial viability, were major obstacles to profitability, and only the collapse of 1903 could convince him of the need for change. His reputation as one of the most eminent British electrical engineers was, however, untarnished by these events and, as we shall see in the next chapter, between 1904 and 1914 Ferranti was able to exploit his position by raising support for a variety of projects. It was in this period that several honours were bestowed upon him for his services to the industry, and clearly his contribution to the development of power-station practice was still recognised by contemporaries as one of the most crucial in the short history of electrical engineering.

4

The freelance engineer

No longer directly involved in the management of Ferranti Limited, in the years leading up to the First World War Ferranti enjoyed a fruitful period exploiting both his patents and his reputation in the electrical industry. He was to take out a total of fifty-eight patents between 1903 and 1914, covering such diverse fields as turbines, steam valves, textile machinery, rubber tyres, gyroscopes and electricity meters. To examine his career in this period is to gain some insight into the world of the freelance engineer, expanding our study of the unresponsive nature of the electrical market into a more general analysis of the response to new ideas in Britain. When invited to give the James Watt Lecture in 1913, Ferranti argued that one of Watt's greatest discoveries had been Matthew Boulton, the 'courageous capitalist' who had financed the production of his steam engine. In that same speech, he pleaded for more 'Matthew Boultons' to come forward, a plea that came from the heart of an engineer who had spent a lifetime canvassing support for his schemes. It will become clear from what we shall see in this chapter that the British scene was uneven in this respect, and some of Ferranti's most innovative work was not always rewarded either with adequate financial support or with significant commercial success. At the same time, Ferranti Limited recovered steadily from its earlier embarrassment and, while Tait, Whittaker and Anderson were responsible for creating and implementing an ambitious market-led strategy, any success enjoyed can partly be attributed to Ferranti's work in developing a highly competitive range of meters, switchgear and instruments. He continued to make some technical input, especially into the Meter Department, but his greatest contribution to the company's prospects after 1903 came in his attempts to improve the degree of cohesion within the electrical industry. Up to 1910, despite the creation of two trade associations, effective unity had been a myth among electrical manufacturers but, by assisting both in the creation of a more powerful trade association and in the reform of the IEE, Ferranti facilitated the process of stabilising conditions within the industry. In his widely-publicised presidential address to the IEE in 1910, he also made another impassioned demand for a radical overhaul of the electricity supply industry. This he regarded as an essential component in Britain's drive for greater efficiency and improved economic growth in the face of powerful foreign competition, indicating that he never lost sight of his life-long dream of seeing the creation of an 'All Electric' age. Overall, it is a peculiar period in his career, packed with interesting developments, and it was only brought to an end by the tragic events of 1914, the year in which he once again took over the helm at the Crown Works. This return to executive management was to mark the beginning of his final career phase, a development covered in the next chapter, but in the years 1903–14 he was to enjoy the freedom of operating away from the confines of a manufacturing business. Ironically, perhaps, he was to enjoy a much improved standard of living as a freelance engineer, but this never dulled his interest in technical innovation and engineering. He always remained dedicated to his first love, the application of technology, and not even the acquisition of his own country house could distract him away from this work.

While Ferranti Limited was entering its receivership in the summer of 1903, Ferranti had already started negotiations with Babcock & Wilcox and J. & P. Coats Limited to take up, respectively, his turbine and textile machinery patents. The work on these two projects, as we saw in the last chapter, had been one of the reasons behind the company's financial problems and, after the receiver-managers were appointed in September, it became increasingly apparent that Ferranti would not be allowed to continue the experiments in the Crown Works. Unfortunately, Babcock & Wilcox pulled out of the discussions very early, but J. & P. Coats were much more interested in his ideas on cotton-spinning and doubling. Ferranti had been introduced to this leading cotton-thread manufacturer by two paper-makers of Linwood, Renfrewshire, Robert and William Watson. They had developed a close friendship with him as major suppliers to the British Insulated Wire Company. In fact, in 1904 the Watsons appear to have acted as Ferranti's 'Matthew Boultons', lending him £5,500, interest-free, to purchase all his turbine and textile machinery patents from Ferranti Limited, as well as providing the initial contact with J. & P. Coats, their near neighbours at Paisley. Although this first meeting in May 1903 did not result in a deal, Ernest Coats was sufficiently interested in the patents to continue negotiations. Progress was hampered by the legal difficulties of reassigning these patents to Ferranti early in 1904, but on 10 May J. & P. Coats finally came to an agreement which gave him £5,000 to finance the development of new spinning and doubling machines. The company also provided £1,000 for work on the turbine, but only on condition that another backer would be found for this project, and by August the electrical division of Vickers at Sheffield had agreed to take it over. All this represents even further evidence of Ferranti's ability to raise support from prominent businessmen, an asset which had already been a feature of his career since establishing the first business in 1882. After 1904 he was financed by two of the largest manufacturing companies in Britain and, in the pursuit of such highly diverse engineering projects, he proceeded to make some interesting contributions to current technology.

An essential feature of Ferranti's work on turbines and textile machinery was the creation of an engineering team to supplement his own technical knowledge and share the burden of design and development. At Deptford, as we saw in Chapter 2, he had first assembled a group of engineers and draughtsmen and, although this was disbanded after 1891, by the time of the move to Hollinwood an experimental team had been established which was to prove useful in assisting in the work on the steam alternator. In 1902, when Ferranti had started his turbine project, this team was joined by Albert Hall, an expert on thermodynamics who was to act as a valuable source of expertise on the more theoretical aspects of the experiments in which they became engaged over the next three decades. When Ferranti was pressed to leave the Crown Works from the end of 1903, Hall became his team-leader, commuting between Paisley and Sheffield to supervise the progress on each of Ferranti's main projects. Other important members of the team were A. Tucker, who later made a significant contribution to Ferranti transformer technology, and R. H. Parsons, the cousin of a leading turbine engineer of this period, C. A. Parsons. Over the period 1904–14, these men, never numbering more than six, worked continuously on a variety of different schemes. The textile machinery and turbine projects were obviously the more time-consuming, but Ferranti also directed their attentions to work on such areas as rubber tyres, stop-valves and gyroscopes. The team became something of a professional engineering consultancy and, indeed, some members later ventured off into that field, but most remained with Ferranti for the rest of his life. This is an indication both of the loyalty he inspired in those he worked closely with over the years and of his emphasis on team-work to solve major engineering problems. After 1914 he instilled this approach to product and manufacturing development into the structure of Ferranti Limited, laying the basis for a diversification strategy which successfully exploited the variety of opportunities created by the events of the inter-war period.

Ferranti, of course, had always been keen to diversify his company's product base, but his decision to move into the production of textile machinery in 1902 was probably one of the most bizarre ventures during his period as a manufacturer. He later stated in an application to have his 1902 patents renewed in 1917, after moving into the centre of the British cotton industry in Oldham, 'I had continually under consideration the question of how electricity could be used to improve or speed-up textile processes.' A Platt spinning frame and a Dobson & Barlow doubling frame were installed in the Crown Works, and electric motors were developed to increase the speed at which these processes were undertaken. Ferranti intended to raise the speed of these machines from the conventional 5,000 revolutions per minute (rpm) up to 30,000 rpm, but although 15,000 rpm were obtained with electric motors it was decided by 1903 to replace them with air-driven turbines. He also dropped the idea of working on spinning frames, because the British cotton industry was largely dependent on the intermittent mule spinner, while his designs were more appropriate to the continuous ring machine. In this context, it is surprising that Ferranti never approached one of the large textile machine manufacturers in his area, like Platt's of Oldham, because they were major exporters of ring-spinning machines. There is no evidence of any negotiations with such companies and one can only conclude that Ferranti concentrated on his links with J. & P. Coats because the financial support seemed more assured. Nevertheless, as we shall see, this proved to be a barren path to follow, especially when he agreed to devote his attention to wet-doubling. This process was used throughout J. & P. Coats's cotton thread mills and was by far the most complicated, compared to spinning. Ponting (1973), who has described these experiments in great detail, is certainly of the opinion that the move into wet-doubling created some of the more obstinate problems. He also questions J. & P. Coats's decision to support this project, an important point which deserves close analysis.

Formed in 1890, with a nominal capital of £5,750,000, J. & P. Coats Limited was the strongest of the large textile conglomerates created in that decade, accounting for approximately eighty-five per cent of British cotton-thread production. Owned and managed by the enterprising Coats family of Paisley, in association with the talented businessman O.E. Philippi, the company was the third largest manufacturer in Britain by 1905, with an estimated market value of £11·2 million. In contrast to many British textile companies at this time, they were willing to invest heavily in ambitious development programmes to retain their market lead. The company was particularly famous for the 'Ferguslie System', by which the efficient organisation of their large works was maintained, and the decision to sponsor Ferranti's ideas is a further indication of this forward-looking approach. It was, in fact, Ernest Coats who was responsible for this decision, being the Director in charge of production and a recognised expert on thread-doubling, but the failure of this particular project inevitably casts doubt on his wisdom. Of course, in the pursuit of technical excellence, some waste is inevitable, but Ponting (1973) argues that this project, which was actually to continue until 1931, contained insurmountable problems an expert might well have predicted. If Ferranti is to be blamed for his failure to solve all these problems, then J. & P. Coats must share some of this blame, while unforeseen external factors may also have contributed to the eventual outcome.

The main difficulty with which Ferranti had to cope, after settling on the use of air turbines to drive the machinery, was the dynamic balancing of the tubular flyers as they spun round at those unprecedented speeds. Although he made imaginative use of air, instead of oil, bearings, and even experimented with such materials as chromium and carbon in the manufacture of the most vulnerable components, by 1914 progress had been slow, despite the expenditure of over £52,000. In the inter-war period, further work was undertaken and in 1927 a bulk experiment using eight wet-doubling machines was attempted at J. & P. Coats's Ferguslie Thread Works, but Ferranti could still only report that more work was required. It was calculated, rather cruelly, that a 10,000 spindle Ferranti machine would require

10,000 machinists to supervise production, such was the complexity of the unit. In 1932, after Ferranti's death, even Ernest Coats was forced to admit that the project had been 'an utter failure', and a total of approximately £120,000 had been spent in the development of a machine incapable of making any impact on J. & P. Coats's production costs. It is important to stress, however, that certain trends in the cotton industry played a role in undermining the economic rationale behind Ferranti's design. This was discovered in the 1927 bulk experiment, when it became apparent that, even if they could be coaxed into performing consistently, the output per square foot of mill floor was very much higher in the Ferranti machine than in the conventional doubling process, but so was the cost of labour, because the former made six ounces of yarn quickly, the latter made one and a half pounds more slowly. In a period when great emphasis was placed on the reduction of labour costs, the Ferranti machine was simply uneconomic, with its high removal and loading costs resulting in an increase in the cost per unit of output. On top of its undoubted mechanical inconsistency, this served to confirm Ernest Coats's opinion that the company had failed to sponsor the right project.

That Ferranti's textile experiments never came to anything may be regarded as a damning indictment both of his technical ability and of J. & P. Coats's foresight as experts in the field. These may be harsh judgements, particularly in the light of Ponting's (1973) conclusion that the British cotton industry's 'historical tendency ... to be most conservative towards innovation' may have played an important part in the eventual outcome. Ponting, in fact, argues that this feature of the scene was not vital, but a deeper examination may lead us to feel that Ferranti was forced into a corner by the traditional reluctance of British cotton-masters, especially in the Oldham vicinity, to introduce more efficient spinning machinery. After all, his assistant at Paisley, A. W. Edwards, argued that Ferranti's designs would have been more appropriate to continuous-ring spinning, but over eighty per cent of the sixty million spindles in operation in Britain by 1914 were of the intermittent mule variety. There were also few cotton companies in Lancashire with the resources to sponsor such expensive development projects, and clearly Ferranti was pushed towards J. & P. Coats, with their more sophisticated wet-doubling process. Ponting argues that had the company been in a more competitive trade, and not controlled eighty-five per cent of the British cotton-thread market, they would have dropped Ferranti's scheme much sooner. On the other hand, Ferranti completely dominated Ernest Coats and, ultimately, it was for this reason that the project continued for so long. Unfortunately, neither could make it into a success, and the textile machinery work remains as yet another example of how Ferranti's mind wrestled with the more interesting mechnical problems of his day.

One of the key mechanical developments of the period leading up to the First World War was the introduction of the turbine, and Ferranti was particularly interested in this new prime mover for two reasons: in the first place, it jeopordised his steam-alternator business but, at the same time, he was convinced that the turbine would significantly improve the prospects for electricity by reducing fuel costs. Ever since Charles A. Parsons had lodged his momentous patent in 1884 for an impulse-type turbine, this machine had been growing in popularity as he steadily improved its efficiency by reducing the steam consumption required per horse power. One of his most successful early installations was at Portsmouth in 1894, where a small 150 kW Parsons turbo-alternator ran side-by-side with the Ferranti steam alternators. This persuaded Ferranti that the reciprocating engine had reached the peak of its development and, in his 1894 speech to the Royal Scottish Society of Arts, he argued that the turbine would soon replace it when the British electricity supply industry started to expand. He even patented an impulse-type gas turbine in 1895, but the pressure of business prevented further development work. Meanwhile, as Byatt (1979) illustrates, Parsons increased his share of the power-station market to thirteen per cent by the period 1904–8, and Parsons (1939) describes how the efficiency and size of the unit increased to such an extent that by 1908 his firm was capable of supplying turbo-alternators with the unprecedented capacity of 6,000 kW. At the same time, other electrical manufacturers were either purchasing licences from Parsons or introducing foreign designs in their attempt to keep in touch with this vital technological development. Quite typically, Ferranti decided to produce his own turbine design. To this end, dispensing with his 1895 idea for a gas turbine, in 1902 he patented the basic design to which he worked over the next twelve years, and purchased a 2,000 kW Parsons turbine for some experiments in the Crown Works. This started him off on a path which resulted in a major improvement in turbine efficiency, although here again market conditions were to hold up greater commercial success.

Ferranti's basic idea for improving turbine efficiency revolved around the premise that the higher the temperature of the working fluid, the greater is the economy to be obtained. To achieve this increase in temperature, he included a superheater in his design, a device Parsons had tried, unsuccessfully, in 1892 when installing three 100 kW turbo-alternators in the Cambridge power station. Ferranti, on the other hand, was confident of his ability to overcome the mechanical problems experienced when working at such high temperatures, and this indeed, is where he succeeded, while Parsons concentrated on more conventional lines. In his turbine, Ferranti superheated the steam up to 420°C before it passed through the first stage and then reheated it between each successive stage, transferring the steam at the end of expansion to the feed water by means of a regenerator in order not to waste any energy. This was further improved in 1906 when he patented the process of 'bleeder heating', by which the boiler would have to give less heat to a pound of steam but the work obtained would not be reduced in anything like the same proportion. Together with the 1902 patent, and the fifteen other related patents he took out in the intervening period, this put Ferranti ahead of the field in terms of ideas, but their practical implemenation involved much more experimental work.

When Ferranti started his experiments in 1902, he could not have envisaged just how much it would cost to produce a satisfactory solution to the mechanical problems involved in such a project. By March 1904, a 1,000 hp turbine had been constructed in the Crown Works to his patents, and in the negotiations with Vickers he informed the Managing Director of their electrical division, Douglas Vickers: 'I think that your risk in the matter would be extremely small. I am satisfied that there is little doubt as to the success of the resuperheating turbine.' Such was the interest in turbines by this time that Douglas Vickers was only too keen to support Ferranti's development work, and he made only a cursory review of the mechanical design. Vickers, the fifth largest manufacturing company in 1905, with a market capitalisation of £7·5 million, were undergoing a process of diversification after the decline in their traditional munitions business in the early 1900s, and the opportunity to acquire what seemed to be some of the most advanced turbine technology around coincided well with this strategy. Consequently, in August 1904 the company agreed to provide £6,000 for a series of experiments to be carried out in their River Don works at Sheffield. Ferranti, indeed, was so confident that he anticipated the eventual creation of a specialist turbine company in association with Vickers, but such hopes were to be dashed not only by the mechanical details of the work, but also because they were to fall foul of the British market's conservatism.

Ferranti seems to have spent much more time on his turbine work than his textile machines in the years up to 1914. This can be confirmed by looking through his sketch-books and private notes, and by noting that between 1902 and 1913 he took out thirty-two patents relating directly to turbines, compared to only nine on cotton-spinning and doubling. With the adoption of temperatures in excess of 400°C, his mechanical skills were inevitably to be tested to the full, not least in the form of blade required. One of the vital mechanical details of a turbine is the composition and placing of the blades around the rim of each separate stage. It is essential that they be firmly attached, because in the early experiments blades frequently became detached from the wheels and only the development of an automatic electric welding machine by Ferranti overcame this troublesome feature of the work. He also produced a novel composition for the blades, protecting the soft steel body with an outer coating of nickel, and by the end of 1905 success seemed to be assured in this area of the machine. This, however, was only one aspect of a highly complicated project, and Ferranti and his team were obliged to continue the development work at Sheffield for almost eight years. Douglas Vickers, like Ernest Coats, proved sufficiently amenable to Ferranti's persuasive abilities, and approximately £50,000 was provided for the experiments, but in 1912, with the expansion of the munitions business, the company withdrew its support. This decision is not surprising because by 1912 only two turbines had been produced and both were used internally by Vickers. On the face of it, Ferranti had failed to produce a commercial product and, although the mechanical design had been perfected by 1912, the market was not yet ready for such a machine.

Vickers' decision to drop the turbine project seems highly rational, particularly in view of the resurgence in their traditional armaments markets, and only with excessive hindsight can one question their failure to persevere with the Ferranti design. They had already spent £50,000 and, although the turbine was ready for sale by 1912, in April 1913 the *Engineer* was only able to say that 'a machine built somewhat on the lines of Mr. Ferranti's turbine will have its day'. The Admiralty, Yarrow's and the French engineering company Schneider were also testing the turbine, illustrating the wide interest aroused by

Ferranti's design. Unfortunately, however, the British electricity supply industry showed no tendency to imitate this interest because, quite simply, the resuperheated turbines was too powerful for their requirements. Even though the scale of British power stations had increased since the mid-1890s, as we shall see in the next chapter, by 1914 the conventional Parsons turbine was still powerful enough for their needs. It was only in 1919 that the resuperheated turbine became accepted practice in Britain, after the firm partly created by Vickers, the Metropolitan-Vickers Electrical Manufacturing Company, produced a 10,000 kW resuperheated turbo-alternator for the North Tees power station of the Newcastle Electricity Supply Company. Ferranti received £2,000 in royalties for this machine, but by the time his British patents had expired in 1921 no further sales had been made. In the USA, where his patents lasted until 1925, General Electric actually purchased the right to build Ferranti's resuperheating turbine for $150,000, illustrating the high value put on the engineering achievement by the inter-war period. This was a fine reward for his work on turbines, but Vickers's enteprise went largely unrewarded, with the exception of the two machines used internally and the North Tees order. Still, they were correct in their decision to write off the development expenditure and exploit the more lucrative armaments market in 1912 because the British power station market was not then ready for such a leap forward in generating capacity. Ferranti had pioneered one of the most important developments in turbine design, and Parsons later admitted this in a speech to the first World Power Conference in 1924, but Vickers never benefited from their enterprise or Ferranti's engineering flair. It was a project which highlighted the restrictive nature of the British environment in terms of encouraging enterprise and innovation, and in this context there was little inducement for Vickers to persevere with such a high-risk product. As Irving (1975) also argues, the profitability of armaments contracts naturally dictated that the leading engineering companies should commit only a small proportion of their resources to new areas like motor-car production or, indeed, to resuperheating turbines. Such considerations, in association with the unfavourable market conditions, were largely responsible for the slow rate of growth experienced in the developing areas of engineering up to 1914 and, although Ferranti and Vickers were responsible for a significant contribution to turbine technology, its commercial impact did not stimulate the creation of the kind of business initially anticipated. Of course, Ferranti earned large royalties in the 1920s from the American users of his design, but he had originally intended to create a specialist turbine company and the failure to fulfil this aim served to confirm his view that conditions were very different on either side of the Atlantic.

Although the turbine project did not achieve its commercial expectations, in connection with this work Ferranti was responsible for another technical development which actually provided the basis for a highly successful business. Working with such high pressures on the resuperheating turbine, the team discovered that the conventional stop-value was inefficient and Ferranti naturally set about devising a suitable alternative. This new design involved the use of a much smaller orifice to convert the pressure of the steam into velocity and, although his team nicknamed it 'Ferranti's Folly' during tests in 1906, he confounded all criticism by showing that his stop-value could improve the capacity of a steam engine by almost one-third. *Engineering* reported in June 1906 that it was a radical departure in valve construction, but at first the principal stop-valve manufacturers, J. Hopkinson & Company of Huddersfield, were so sceptical about Ferranti's patent that they refused to purchase it outright at his offering price of £2,000. Instead, they decided to pursue the less risky policy of paying a 7·5 per cent royalty on all sales, a decision they were later to regret. The Hopkinson agreement was signed in May 1906, and although it took almost a year before the first major order was secured, from the Sunderland Electricity Supply Company, in the lifetime of the patent to 1921 Ferranti received almost £32,000 in royalties. This amounts to a business worth approximately £425,000, on top of which two European companies, Franco Tosi of Milan and Schaffer-Budenburg of Magdeburg, also negotiated the rights to manufacture the Ferranti-Hopkinson valve, Ferranti being paid a further £1,500 on these accounts. Hopkinsons clearly benefited greatly from their arrangement with Ferranti, confirming their position as market leaders in this field. Ferranti's engineering talents had resulted in another important contribution to steam engine technology, albeit one of a less spectacular nature than the work on resuperheating turbines. Overall, indeed, Ferranti's work with Vickers had resulted in some remarkable engineering developments, bringing him great personal satisfaction, and the commercial success enjoyed with the stop-valve would have provided some compensation for the failure to commercialise the turbine patents in the way anticipated in 1904.

As the annual income from the stop-valve patent slowly built up, from £9 in 1906 to almost £3,000 by 1913, the Ferrantis were able to enjoy an unprecedented period of financial stability. Ferranti also drew a salary of £500 per annum from Ferranti Limited for his services as a Director and, as we shall see later, he earned large fees as a parliamentary advocate for electricity supply companies, banishing the financial problems of the 1880s and 1890s firmly into the past. Certainly, he needed a good income to support his growing family, which by 1913 included three sons and four daughters. (His mother and father had died in 1906 and 1903 respectively.) Ferranti later claimed that the Hopkinson royalties paid for his two eldest sons' education at Repton between 1907 and 1912, but this income must have been more important than that because by 1913 he had purchased his own country house, Baslow Hall, near Chatsworth. This illustrates his financial status by that time, but one must emphasise that living in the country did not dull Ferranti's enthusiasm for engineering and industrial affairs. Not for him the life of what Wiener (1981) calls the 'gentrified' businessman, and he also ensured that his two eldest sons, Basil and Vincent, were inculcated with his love of technology. Even at Repton, they were encouraged to read books on engineering and use the school's workshop to improve their understanding of the subject. In 1910, Basil was also sent to Sheffield University on a two-year course in mechanical engineering, thereafter spending a further two years as an apprentice at Yarrow's, the Clyde shipbuilders. In 1912, Vincent followed his brother there to complete a similar period of training. Clearly, Ferranti was guiding them along a very obvious path, driving the engineering ethos firmly into their minds and, as we shall see in the next chapters, this was to pay handsome dividends in the decades ahead. At the same time, by 1910 he had bought them each a motor-bike, perpetuating what Gertrude (1932) remembers as something of a family craze in what she calls 'those rather happy-go-lucky' days. It is clear from the family correspondence that they enjoyed a happy family atmosphere, and only the political and military events of 1914 upset the situation. Both Basil and Vincent were eager to join the Armed Forces in 1914 and they were successful in securing commissions in the Royal Artillery and Royal Engineers respectively, but, as we shall see later, tragedy was to strike the family during the War and Basil never returned from the battlefields of Europe.

One should never gain the impression that Ferranti's new-found prosperity was the result of sitting back and waiting for the Hopkinson royalties to roll in. Apart from his time-consuming experiments in Sheffield and Paisley, and the work with certain electrical organisations, he was heavily involved in two other engineering projects during this period. These projects were related to motor cars and aircraft, and illustrate again his interest in the more fascinating engineering developments of his day. Ferranti had been introduced to the automobile in 1906, at the Paris Motor Show, and he was immeditately struck by the inefficiency of its tyres. One of the major problems for the early car driver on those poor roads was the weak construction of rubber tyres, and Ferranti decided to develop a more durable design based on the idea of pouring molten rubber onto a backing of woven fabric. This was one of the earliest reinforced tyres ever made, and such was Ferranti's confidence in this 'fibre-tread' design that in November 1907 he approached Dunlop with a proposal to test it against their models. The company agreed to finance the venture but it was only by March 1909 that a series of experiments had been carried out at their Birmingham plant. No figures are available on the cost of these trials but it is clear that Dunlop were not convinced that Ferranti's tyres would improve significantly on their solid rubber model. Ferranti complained that they were comparing the performance of his design with the best ever produced by a Dunlop tyre, while, in his opinion, the company's experimental department seemed to be run in a most casual manner. This confirmed his suspicions that Dunlop were more interested in their own development work and, in July 1909, the fibre-tread experiments were terminated by mutual agreement. Another consideration affecting Dunlop's view of the reinforced tyre was the growing number of competitors in this particular market. In fact, after the acquisition of the Helsby Tyre Company in 1902, by 1908 the British Insulated Wire Company were manufacturing a tyre similar to Ferranti's and Dunlop would have had to invest heavily in a development programme to catch up with them. For Ferranti, this created an unethical situation, whereby he was still on British Insulated Wire's Board but working for a direct competitor. He may even have approached the company with his own tyre patent, although there is no evidence to support the point. Whatever the case, in 1909 he had resigned from British Insulated Wire's Board, bringing to an end a long and successful relationship which had brought tremendous mutual benefits.

The fibre-tread tyre incident illustrates the diversity of Ferranti's technical interests in this period, a point further confirmed when we consider his attempt to develop a gyrostabilised platform for the aeroplane. As part of their diversification strategy, Vickers had sent Ferranti to the Grande Semaine d'Aviation de la Champagne at Rheims in 1909. No subsequent report exists, but from his sketchbooks it is clear that Ferranti was particularly interested in the primitive methods employed to stabilise aircraft in flight. This led him to devise a gyrostabilised platform, putting him right in the vanguard of aviation technology. He was, of course, by no means the first to propose the use of gyroscopes in this way because, ever since the late 1890s, various European and American engineers had been experimenting with techniques using these instruments. The most famous, and most successful, of these engineers was the American, Elmer Sperry. He started his work on gyroscope stabilisation in 1907 and, according to Hughes (1971), by 1912, with the extensive support of the US Navy, methods of automatically steadying both ships and aeroplanes had been perfected. Sperry was also able to interest several New York businessmen in his ideas, and in 1913 the Sperry Gyroscope Company was created to exploit what was the most advanced gyroscopic equipment in the world. The *Engineer* in May 1913 reported at length on the vital importance of Sperry's work, but this article prompted Ferranti to write a fierce critique, explaining how in 1909 he had patented a highly effective method of stabilising aeroplanes using gyroscopes working in tandem which differed only slightly from that developed by Sperry. He also noted in 1911: 'when I met Mr. Sperry in London I explained to him the importance of the use of two gyroscopes suitably coupled in any airplane stabilising'. This may well have been crucial to the American's work, because it was only in 1912, at the Bezons trials held by the Aero Club of France, that Sperry first exhibited his aeroplane stabiliser with its two gyroscopes. It is, however, difficult to confirm this picture of events, although Ferranti had clearly been thinking along the same lines as Sperry.

One point Ferranti made in that letter to the *Engineer* deserves much closer attention in the light of his views on the nature of commercial conditions on either side of the Atlantic. While he admitted that other projects had distracted his attentions from the work on gyroscopic stabilisation, in his efforts to generate some interest in the patent he had come to the conclusion that 'in this country one cannot readily find support for new ideas, whereas in the States so far as one can see this does not appear to be very difficult'. He discussed the merits of his ideas with such authorities as F. W. Lanchester and A. A. Campbell-Swinton, but the contemporary view that pilot skill was the only effective method of stabilising aeroplanes was just too strong in England. After the success achieved by Sperry in 1912, in 1914 Ferranti was able to persuade the Royal Aircraft Establishment at Farnborough to test the equipment, but even here attitudes proved little more encouraging. Their test engineers were evidently impressed with the working of gyroscopes in pairs, but concluded: 'the amount of "working-out" to be done at present would appear to make it impracticable unless very considerable sums of money are to be sunk in experiments on this line alone'. This confirmed the views stated in Ferranti's 1913 letter to the *Engineer*, and, while Sperry had been supported by both the US Navy and private financiers, Ferranti's ideas were largely ignored, giving the Americans a lead in what became another vital new sector of the engineering industry in the twentieth century.

In view of his experiences in the years 1903–14, Ferranti seems well-qualified to comment on the propensity to support new ideas in Britain. Of course, J. & P. Coats, Vickers and Dunlop had provided the finance for certain projects, illustrating how some of the larger manufacturers were willing to engage in risky ventures. Hopkinson's had also taken up the manufacture of his radical new stop-valve, albeit at little risk to themselves. Yet even where the mechanical problems could be overcome, for example in the resuperheating turbine, the market conditions failed to encourage greater enterprise. One might also add that his three main backers were quick to drop his schemes once it became apparent that their strategies diverged from the market logic behind Ferranti's ideas. This was only rational on the part of the companies, but the gyroscope episode came at the end of a period in which, with the exception of the stop-valve, a combination of factors had prevented the more extensive commercialisation of Ferranti's patents. One must naturally question the viability of some of these patents, especially with regard to the doubling machine and rubber tyres, but, in the case of the resuperheating turbine and the gyroscope, greater effort would have put Britain at the head of the field. When one combines these experiences with the story of the electrical industry's development since the early 1880s, it is difficult to escape the conclusion that entrepreneurs were faced with some insurmountable problems in establishing the commercial viability of new technology in Britain prior to 1914. Ferranti was right when he asked for more 'Matthew Boultons' to come forward, in his James Watt Lecture of 1913, but this in itself would only go part of the way towards solving the real problems. What Britain really required was a responsive market for new products because, once this appeared, the 'Matthew Boultons' could be persuaded to support ventures with the inducement of more secure returns. In the USA, financiers were much more willing to invest in the electrical and aircraft industries, with their expanding markets, and this attraction was the vital difference between American and British conditions. Ferranti struggled in vain against these obstacles, only succeeding in making one significant commercial success, but it is difficult to see just how he or his supporters could have achieved more in the context of Edwardian Britain's generally unresponsive environment for new technology.

While Ferranti was engaged on these various engineering projects, he also kept a close interest in the progress of the electrical industry. He was still a Director of Ferranti Limited and the 'All Electric' ideal was never very far from his mind, influencing both his work on the turbine and the efforts he made to stabilise conditions for electrical manufacturers. Since his company's demise in 1903, the state of the electrical market had deteriorated even further, affecting both the older British manufacturers and the much larger new generation of businesses established around the turn of the century. In fact, because the former had rationalised their product ranges earlier in the decade, it was the larger companies which suffered most. The largest of the new generation, British-Westinghouse, was even forced to write off half of its equity capital in 1905, and survived only on the basis of financial support provided by its American parent company. This situation was caused partly by over-optimistic market projections in the boom of 1896–1903, but mainly by the severe price competition which resulted in price reductions of up to forty per cent by 1907. Ironically, as Byatt (1979) notes, this was not due to a fall in demand, because total domestic production of all electrical machinery (at 1907 prices) actually increased from an average of just over £2·8 million per annum in the period 1902–4 to around £4·6 million in the years 1908–10. The principal problems seem to have been a combination of overcapacity, a general reluctance to engage in discussions on prices, and the power vested in the obstructive local authorities which prevented greater standardisation of production. Such was the unremunerative nature of the electrical business that by 1910 almost 65 per cent of the capital in electrical engineering was not producing a return, according to the Monopolies and Restrictive Practices Commission (1957), and the *Electrical Review* of February 1910 reported: 'cut-throat competition prevails in almost all branches of the industry'.

Having reconstructed Ferranti Limited in 1905, Tait and Whittaker would have viewed the deteriorating market conditions with alarm. Although they had rationalised the product range by closing the big loss-making steam-alternator business, and had streamlined the financial structure, they would require favourable prices in order to pay the shareholders adequate dividends to compensate them for their patience. As in the past, it was the meter on which so much depended in the period 1905–14 and, in order to secure a good return on this business, they embarked on a pioneering policy of price-stabilisation. In fact, the key figure in these negotiations was A. B. Anderson. He had been retained by Tait and Whittaker because of his undoubted financial and business talents, and by 1908 he was appointed Managing Director in recognition of his vital contribution to the company's progress. Anderson was also to become one of the principal architects of a powerful new trade association, after playing a similar role in the meter trade. As early as 1906, he had negotiated a mutual interchange of meter patents with Ferranti Limited's major rival in the DC meter market, Chamberlain & Hookham, and by the January of the following year a detailed price list had been introduced. This arrangement was later extended to cover their largest export markets, in India, Burma and Australia, guaranteeing stable prices for what was the company's main product. By 1910, even the formerly obstructive BTH had agreed to join the ring, and thereafter the Meter Department made an excellent return on sales. The AC meter market was rather more difficult to organise, because British-Westinghouse refused to abide by any regulations on prices, but even here by 1912 agreement had been reached, bringing stability to both the DC and AC sides of the business. (It is evident here that the two main American subsidiaries were a disruptive influence in the efforts to stabilise prices, with their parent companies hoping that fierce competition might drive out the rivals, but, as we shall see later, after 1911 their attitudes were to change.)

Although price stability was important to the company's recovery after 1905, this was only one part of Anderson's strategy to expand sales and profits. The home market for meters was highly seasonal, with the electricity supply stations advertising for tenders to supply the bulk of their requirements in the autumn. This business might keep the Crown Works busy over the winter months but, in order to maintain production levels throughout the year, Anderson recognized that an aggressive marketing policy in both home and export markets would be required. He had reported to a Board meeting in October 1906: 'the prospect of obtaining the necessary turnover at remunerative prices is such as to necessitate the consideration of some more drastic steps than merely quoting and following tenders'. Tait and Whittaker agreed with this and over the next six years an extensive network of salesmen and agencies was established. At home, by October 1907 every major town and city was covered by the company, and new London offices in Central House, Kingsway, were taken in 1912 to house the newly-created Sales Department. Anderson also travelled abroad to find suitable agents in Europe, South Africa and North America, while by 1912 three subsidiary marketing companies had been created, in Germany, Austria and Canada, to boost sales. The Austrian and German companies were never very successful, but the Canadian operation, as we shall see in the next chapter, eventually blossomed into a major manufacturing, assembly and sales company in its own right. In spite of the strong American influence in the Canadian market, the Ferranti Electrical Company of Canada Limited, formed in 1912, became a vital component in the company's drive to expand sales in order to spread meter production throughout the year.

Anderson's aggressive marketing strategy was a new departure for the company, but it is important to note that he was by no means an innovator among electrical manufacturers in this respect. All the major new generation companies had established extensive marketing and sales networks in the home and export markets, contradicting what has in the past been a general criticism of British businessmen in their alleged indifference towards such managerial tools prior to 1914. Ferranti Limited were not to be left out in this race for business, and every effort was made to establish contacts with customers in a large number of countries. This proved successful in boosting sales, but a closer examination of the agencies' geographical location will reveal some weaknesses in the approach of British electrical companies. In fact, most of the agencies were concentrated either in the Empire or in such developing economies as South America and the Far East, whereas the larger, more sophisticated markets were to be found in Europe and the USA. This added further to the relative backwardness of the British industry, because the agencies only stimulated the production of less advanced products, while German and American conglomerates could exploit the demand for new technologies in their heavily protected home markets. Prominent electrical manufacturers like Hugo Hirst of GEC and G. Flett of Dick Kerr played an active role in the protectionist movement created by Joseph Chamberlain after 1903, but the British electorate preferred free trade and no effective changes in commercial policy were introduced until after the Great Crash of 1931. For Ferranti Limited, Anderson was involved in drumming up local support for Chamberlain's plans, becoming President of the Failsworth and District branch of the Tariff Reform League, and in the inter-war era more effort was put into the cause of creating a guaranteed home market for the industry, but the government remained unmoved. It was regarded by most electrical manufacturers as essential that foreign competition should be excluded from the British market but, in the meantime, they were forced to concentrate their export efforts on the Empire and developing economies to supplement the home business. For Ferranti Limited, however, the problem of the low level of sophistication in their major export markets was never great. After all, they were mainly engaged in selling electricity meters, rather than large generating units, and the agency network proved to be an excellent vehicle for expanding sales of the company's staple product. At the same time, exports of switchgear and transformers were slow to increase, and it was only in the inter-war period that sales abroad of the latter took off. The Empire remained vitally important to this strategy of expanding exports and, in the years for which we have accurate statistics, 1917–39, Ferranti Limited sent on average eighty per cent of its foreign business to such countries as India, South Africa and Australia.

Anderson's market-led strategy proved to be a great success after 1907 and, in conjunction with price stability in the DC meter market, it provided Ferranti Limited with a solid basis for recovery. As we noted in the last chapter, most DC meters were sold in Britain, while abroad Ferranti Limited were able to sell most of their AC meter output. In the year before the company's collapse in 1903, Ferranti, in association with the Chief Meter Engineer W. Hamilton, had patented a much-improved AC meter, the Type B. This proved especially popular overseas and by 1913 output was worth over £62,000. At the same time, DC meter production increased to £72,500, and the department's profit and loss account showed a surplus of £26,230 on this business. In view of these results, the Meter Department was still keeping the company afloat, as it had done in the 1890s, accounting for over sixty per cent of total sales and most of the profits. This confirmed the efficacy of Anderson's choice of strategy. Ferranti Limited, indeed, were very anxious to push these strategies as one of the more obvious solutions to their earlier predicament. At the same time, they did not ignore the need to make continued improvements to their product range, and the departmental managers were always hired for their engineering flair, rather than for any business acumen they may have possessed. Ferranti Limited still retained their reputation as manufacturers of high quality products, and Ferranti was frequently consulted, at Board level, on the improvements required when competition was affecting sales. This was particularly true with meters, but with regard to switchgear Ferranti assisted the manager, M. B. Field − H. W. Clothier had left in 1904 to join Reyrolle − in the development of more sophisticated switchboards. Some prestigious orders were secured as a result of this attention to design, for example the distribution boards for the two Cunard ships *Mauretania* and *Lusitania*, but the Switchgear Department was never very profitable up to 1914, with competition forcing down prices continually throughout the period.

Anderson's success in stabilising DC meter prices was not only earning Ferranti Limited better profits after 1907, by the end of the decade it was also making his reputation as one of the leading figures in the electrical industry. After the ruinous competition of the last seven years, by 1910 a growing number of contemporaries were coming to recognise the value of closer co-operation. Although the National Electrical Manufacturers Association (NEMA) had been formed in 1902, following the collapse of the ineffective EPMA, little had been achieved in the way of unifying the interests of electrical companies. Since 1902, NEMA had only been successful in negotiating on such matters as railway rates, customs classifications and exhibitions, but it did serve as a meeting-place where mutual problems could be discussed. Ferranti Limited had joined the organisation in 1906, and in 1910 A. B. Anderson was appointed to represent NEMA on a committee created jointly with the IEE to push the idea of an 'Electrical League'. This committee published a pamphlet entitled *The Need of Co-operation in the Electrical Industry*, advocating greater stability among manufacturers and, on the basis of this activity, the *Electrical Review* of October 1910 proposed a Round Table Conference to discuss such matters. It frequently published the views of leading figures in the industry, and Ferranti is quoted as arguing: 'I think the whole tendency of modern business is to avoid excessive competition by some process or other. Combination, reasonably and fairly carried out, seems to be distinctly beneficial ...' This concurred very closely with Anderson's views and together they were working energetically behind the scenes to bring about a decisive change in the industry's view. It was this activity which led ultimately to the formation of the British Electrical and Allied Manufacturers Association (BEAMA) in September 1911, one of the most powerful trade associations of that era. Several other electrical companies were naturally involved in this development, but Ferranti and Anderson seem to have played vital roles, a point well illustrated when one considers that the former became BEAMA's first Chairman, to be succeeded in 1913 by the latter. Of course, one should not ignore the fact that demand was also picking up at a faster rate after 1911, encouraging even the American manufacturers to put aside their natural hostility to price-fixing, but without the work of men like Ferranti and Anderson little progress would have been made in the field of trade unity.

The creation of BEAMA was a major breakthrough for those in the industry who had worked actively for unity. It highlights the importance of Ferranti's indirect means of supporting Ferranti Limited's recovery, as well as the direct effect it was to have on general profitability in the industry. Under Ferranti and Anderson, BEAMA had grouped all the major electrical companies under a common flag, with the 125 members of 1914 including even the American and German subsidiaries. BEAMA was actually run by a full-time Director in this period, D. N. Dunlop, and financed by higher subscriptions than with NEMA, with an administrative bureau to supervise the fourteen sub-committees which covered the major product groups. At Annual General Meetings, it was curtly noted that the proceedings of these sub-committees were not for general publication, because they were responsible for the rampant price-fixing which became the key feature of BEAMA's activities over the next fifty years. Although the organisation was also concerned with co-ordinating the industry's policies and presenting them to politicians at local and national levels, its main contribution came in the area of prices. By stabilising price levels in the home market, BEAMA was able to improve profitability significantly and, in association with the growth in demand after 1911, this was vital for the industry's survival and expansion. For Ferranti Limited, this made no difference in the meter trade because the negotiations with Chamberlain & Hookham since 1906 had already provided remunerative pricing arrangements, but when the Transformer Notification Agreement was first signed in 1913 they were among the five manfacturers responsible for this association. This reflected the company's ardent desire to provide stable conditions in the home market for their principal products and, although no similar developments occurred in the switchgear trade, Anderson ensured that Ferranti Limited participated in those agreements which would further this aim. Ferranti had clearly played an important role in all these negotiations, and his persuasive abilities were used to good effect in bringing most electrical companies into the BEAMA fold. He was undoubtedly a senior statesman in the industry by this time and his position as BEAMA's first Chairman gave him the vantage point from which he could preach the doctrines of unity and stability. Given the rise in membership, and the strength of the price-fixing agreements, evidently most electrical manufacturers were convinced that his opinions were correct, and up to the early 1920s BEAMA exerted a considerable influence in the industry.

Although Ferranti was principally appointed Chairman of BEAMA because of his well-known views on the need for price stability, another important reason was that in 1911 he was serving his second consecutive year as President of the IEE. This gave the impression to any waverers of a closely-knit industry. The IEE itself had also been undergoing a process of change and expansion since the early years of this century, in the first place because it was outgrowing the temporary facilities hired from the Institute of Civil Engineers, but mainly because many of its senior members were convinced that the IEE was losing contact with the practical men in the industry. The problem of space was quite easily solved and, by June 1910, the current premises of the IEE, Savoy Place on the Victoria Embankment, had become its permanent home. When Ferranti was first appointed President in November 1910, a process was by then under way for increasing recruitment by creating examinations for Associate Membership, by which men with a commercial or managerial background could enter the Institution's ranks. It was Ferranti's task as President to oversee the introduction of these examinations, and he travelled around each of the local sections explaining the need to re-equip the IEE with members from all parts of the industry in order to strengthen the profession. This was known as the 'Broader Policy', and such was his success on these tours, and his interesting approach to public-speaking, that he was invited to serve a second term as President, the first time the IEE had bestowed such an honour on any of its leaders. Running parallel to the birth of BEAMA, this is further evidence of his eminent status in the industry. The examination system was also highly successful because, in the year it was first brought into operation in 1913, total membership jumped from 6,600 to 7,084, one of the biggest annual increases ever experienced. Another important innovation at the IEE during Ferranti's tenure of office was the creation of the Industrial Committee. This powerful branch of the organisation, incorporating the old Parliamentary Committee, reflected the greater interest in commercial affairs and, as the *Electrical Review* of April 1912 stated, it is clear that Ferranti 'aroused the Institution to a realisation of its magnificent possibilities' in linking it to the general developments in the industry.

One could not leave this episode in Ferranti's career without mentioning his controversial Presidential Address to the IEE in 1910. This speech, entitled 'Coal Conservation, Home-Grown Food, and the Better Utilisation of Labour', we shall discuss in greater detail in the next chapter, but it was another crucial statement of his hope that an 'All Electric' world would very soon be created. Just as with Deptford, his audience was captivated by the plans to build a national network of power stations supplying electricity and vital coal by-products in order to improve efficiency within the British economy. Although other British engineers, in particular Charles Merz and W. McLellan, had also started developing large-scale, high-tension AC systems by this time, Ferranti's speech was the first public statement demanding the creation of such an ambitious scheme. It was to take a further sixteen years of lobbying before the State accepted the need for such a plan to integrate the electricity supply industry but, in the meantime, Ferranti and his fellow-electrical enthusiasts continued to speak publicly on such matters. In his role as a consulting engineer to the electrical industry in general, Ferranti was also able to pass on the benefit of his ideas and experiences to influential parliamentary committees when hired to give evidence in support of electricity supply bills. His reputation as one of the best-known advocates of large-scale supply systems, and as one of the most eminent men in the industry, earned him almost £3,000 between 1902 and 1911 as a public advocate of such schemes as the extension to the NESCo network and the creation of a Somerset & District Power Company and the North Wales Power Company. It is naturally difficult to estimate just how influential this evidence proved to be, but by stating his firmly-held views in such circles progress was being made in the creation of new types of power companies. We shall discuss these developments in greater detail in the next chapter, but clearly Ferranti's reputation was spreading ever-wider in the years leading up to the First World War as electricity made rapid strides as a viable alternative to the lighting, heating and power markets.

Advocating the wider extension of the electricity supply network was, of course, an excellent method of stimulating demand in the markets for Ferranti Limited's main products, meters, switchgear and transformers. Indeed, the use of electricity was increasing up to 1914, with the number of units sold rising from 180 GWh in 1900 to almost 2,000 GWh in 1913, and, as we saw earlier, meter output benefited from the connection of a larger number of customers to the network. Ferranti Limited even introduced a prepayment meter in 1910, but up to 1914 the conventional meters provided the bulk of the business because it was only in the 1920s that demand for the pay-as-you-burn methods really took off. The most ambitious range of products introduced by the company before the First World War was domestic appliances, in the hope that middle-class customers might purchase these labour-saving devices. The American manufacturers dominated this particular market because demand for electric fires, irons, cookers and even vacuum cleaners had already provided the incentive to invest in expensive development programmes. At the same time, the British market was very conservative, with the traditional coal-fired kitchen ranges proving much more popular, while domestic servants could still be found in large numbers to perform the dirty and laborious tasks. Regardless of these rather negative indicators, by 1912 Ferranti Limited were marketing a wide range of domestic appliances, from electric ovens through to a novel parabolic reflector fire which could be pivoted on its base to act either as a grill or as a fire. Ferranti made an indirect contribution to this venture by convening a conference in 1913 to standardise the general safety rules for electric cookers. This scheme, known as the 'Point Fives', served to introduce Ferranti products to a wider audience, but even such extensive publicity could not make this business profitable. In common with most British manufacturers, their sales were poor, averaging only just over £2,000 per annum up to 1914, and it was only in the 1930s that demand for such products stimulated volume production.

Despite the failure of the domestic appliance business, by 1914 Ferranti Limited was in a much stronger financial position than it had ever been. The cautious financial policies pursued by Tait and Whittaker and the ambitious strategy created by Anderson had ensured that the company would maximise the opportunities available in a difficult period both for them and for the industry in general. Ferranti, too, seems to have played a part in this recovery process, albeit, with the exception of the Type B meter, an indirect part, in working for effective unity among manufacturers. The stability created by the meter agreements and by BEAMA provided the remunerative prices essential for the recovery of the company's finances and, as the appendix reveals, by 1913 net profits were once again reaching appreciable levels. In analysing these figures, one must remember that they are stated after the payment of several major calls on the company's earnings. No dividends were paid to the shareholders throughout this period, but the deadweight of interest payments, averaging £8,000 per annum, hampered the growth in earnings available for internal use. Of course, interest on the £100,000 debentures had to be paid each year, and in 1912 another fixed-interest mortgage of £50,000 in 6% Five Year Notes was taken out to finance expansion of meter production. Parr's Bank also continued to support the company with an overdraft, which varied between £20,000 and £28,000. Other deductions from profits included a generous allowance for depreciation each year, leaving only a small proportion of gross profits for transfer into the General Reserve Account. Nevertheless, by following such cautious policies, the Board was able to finance most of its own requirements, and by 1913 the reserves stood at over £22,000. This is not to deny that the company was still very short of working capital – between 1910 and 1912 one of the Directors, J. M. Henderson, loaned £6,000 on the security of certain book debts to provide much-needed cash while customers' payments remained outstanding – but illiquidity was never as acute as it had been prior to 1903. Naturally, given the losses of approximately £106,000 between 1903 and 1906, and the problems in the electrical industry, Ferranti Limited could never have raised equity capital, but the increase in output to £300,000 by 1913 illustrates the underlying rate of progress since 1905. By the end of this period the Crown Works employed over 1,750 men and women, and a more secure future seemed to be assured. It was on this basis that the company was to expand rapidly in the inter-war period as the market for electrical goods received a series of fillips from both State-induced activity and the inner compulsions of the British economy.

The acquisition of professional businessmen in 1903 had evidently proved to be a critical factor in the company's recovery from its partially self-induced malaise. Tait and Whittaker's financial skills, bolstered by the support of J.M. Henderson, ensured that any available cash was spent prudently. In 1913 another important Director was to join the Board, Major R.W. Cooper. A.W. Tait had built up an extensive range of contacts in the business world by this time and one of his principal directorships outside Ferranti Limited was the British Aluminium Company. He was later to become Chairman of this company, but one of his fellow-directors was Major Cooper and, in bringing this well-connected businessman on to the Board, Ferranti's Chairman was providing the company with a valuable source of advice of financial matters. Major Cooper was actually to remain as an independent director of Ferranti Limited until 1963, providing continuity in the company's direction over a period of rapid change. One must note, however, that, although this infusion of fresh talent facilitated a change in Ferranti Limited's fortunes, it was Anderson's market-led strategy which proved of greatest importance. Anderson's work in establishing Ferranti products in hitherto untried markets, his improvement in the sales effort, his efforts in stabilising meter prices, and his tireless work-rate underpinned Ferranti Limited's success in the years up to 1914. Of course, as we noted earlier, the company was still based largely on the products developed by Ferranti prior to 1903, and the founder deserves credit for his foresight in producing a range which remained competitive for so long, but Anderson's work after 1905 stands out. Such was his importance to the company that Ferranti once introduced Anderson with the compliment: 'He is the company.' Indeed, with internal competition at its most intense between 1903 and 1911, Ferranti Limited might well have suffered as badly as it did prior to 1903 without Anderson's leadership.

This period in Ferranti's career stands out as one of the most intriguing episodes in a life dotted with quite remarkable events. No longer directly concerned with the executive management of Ferranti Limited, yet influential in assisting its progress through his general work with the product range and with BEAMA, he became involved in a series of projects which illustrate both his intellectual diversity and his untiring efforts to bring about the 'All Electric' age. His reputation as a senior statesman in the electrical industry was recognised by his appointment to the Presidency of the IEE for an unprecedented double term, while in 1912 Manchester University conferred upon him an honorary doctorate for his contributions to electrical engineering. Hereafter, we shall refer to him as Dr Ferranti, partly to avoid confusion with the name of his company, but mainly because after 1912 he was always known as 'The Doctor' at the Crown Works. It is important to note that this is the first time he had been given a public award for his electrical work and, although this would never have worried him, his aversion to non-industrial activities probably accounts for this dearth. He never used his reputation to enter politics, either nationally or locally, preferring the life of an engineer to public work. His work with BEAMA and the IEE seem to have been temporary aberrations because at no other time in his life did he become embroiled in industrial politics. After 1914 especially, he was far too busy at the Crown Works, having returned to his business in the wake of the tragic events which occurred in that year. For the last sixteen years of his life, Dr Ferranti was heavily involved in improving the company's product range and enhancing their prospects for future growth, illustrating that he still had much to contribute in establishing the Ferranti name as one of the major electrical businesses.

5

The return from exile

The Chicago economist Joseph Schumpeter has described the first four decades of the twentieth century as a long wave of economic development carried in large part by the electrical revolution. Although in Britain these developments took some years to gather momentum, as we have seen in earlier chapters, in the period 1914–39 decisive changes occurred in the attitudes to, and consequent size of, the electrical industry. Indeed, the inter-war years provided a much more encouraging environment for electrical manufacturers, with the construction of a national grid network, the greater use of electricity in the home and in industry, and the emergence of new products like radio. The establishment of a grid network, in particular, was the achievement of a goal for which Dr Ferranti had been striving since the 1880s, and it was appropriate that his firm should play a vital role in the design and construction of certain parts of the system. Of course, one must never forget that he was more of a voice in what had become a crowd demanding such an innovation in Britain after 1914, but his position as a pioneer of high-tension AC electricity generation and distribution placed him naturally among the leaders of this increasingly powerful movement. The Ferranti company was to benefit enormously from these new market stimuli, with Dr Ferranti returning to the Crown Works in 1914 after almost a decade of exile from Hollinwood. We shall examine in this chapter how he led the company out of the problems experienced in 1914 into a much healthier era of expansion and financial strength, in the first place by converting the Crown Works into a munitions business. The War proved crucial to the British electrical industry on the manufacturing side because it created a valuable breathing space in reducing the threat from foreign competition, and while Ferranti became heavily involved in munitions production Dr Ferranti was able to exploit some of the opportunities available by establishing a business which became one of the company's hallmarks in the next sixty years, the manufacture of large transformers. Ferranti's reputation in this field was eventually to secure profitable orders for the transformers required in the national grid, and the pursuit of ambitious development programmes extended this reputation across the world. Dr Ferranti had always been keen on extensive product development and, in an era when many other businesses were investing heavily in research departments, he used the funds generated from the continuously successful Meter Department to instill his life-long pursuit of technical excellence into the very structure of his company. This approach, along with the growing financial strength of the company, provided the basis for a diversification strategy which took Ferranti into new, rapidly-expanding markets in the 1920s like radio. Dr Ferranti's later years were to be as exciting and innovative as his first decades in business, and they prompted such important changes in the company's product range that by 1939 it bore little resemblance to that of 1914. Free from the burdens of management, he could concentrate on what really appealed to him while more talented businessmen supervised the finances and organised the sales effort. Ferranti had developed into a well-balanced business by the 1920s, and the years 1914–30 were vital in providing the basis for any progress achieved in future decades.

Although Ferranti had made steady progress since 1905, slowly easing themselves out of the earlier financial malaise, the events of 1914 materially affected the company's prospects for faster growth. The most significant event of that year upsetting the electrical market was obviously the onset of the First World War, but the company's response to the consequent disruption was severely restricted by the untimely death of the Managing Director in May. A. B. Anderson had been a frequent visitor to the company's Canadian subsidiary but on a return journey in 1914 his ship, the *Empress of Ireland*, was sunk in a collision with a Norwegian collier while travelling through a bank of fog in the Gulf of St Lawrence. Of the 1,476 passengers and crew on board, 1,014 were lost. Not only was this the biggest maritime accident since the sinking of the *Titanic* in 1912, it also deprived Ferranti of its most dynamic executive. It is clear that his loss, at the age of only 44, was felt throughout the industry, and the BEAMA Council meeting of June 1914 passed a resolution noting 'the untimely death of their late Chairman ... through whose efforts mainly [BEAMA] has reached its present position'. This re-emphasises his vital contribution to the establishment of BEAMA in the debate at the beginning of the decade, while as Ferranti's Managing Director he had been responsible for the aggressive marketing strategy pursued since the company's reconstruction in 1905. He may not have had any ready solutions to the unsettled conditions pertaining in the summer of 1914, but the Board seem to have stumbled along after his death with no clear policy. It would appear that they refused to envisage any radical measures in staving off what was an alarming decline in demand for electrical products, especially from customers abroad. They believed, like most of their contemporaries, that the War would not last long. Dr Ferranti, on the other hand, was not convinced by the claims of the 'Business as Usual' movement and, as soon as War was declared in August, he decided to involve himself in the day-to-day affairs of the company.

Dr Ferranti, of course, had been excluded from taking an active part in business at the Crown Works since 1904, while his other projects had kept him extremely busy. Even his attendance at Board meetings had been sporadic, but in August 1914 he instructed all the senior managers to submit reports on the market and financial prospects for the immediate future. These reports made poor reading and he instructed the Sales Department to circulate to all customers who had formerly purchased German equipment, in the hope that new business might be secured. More importantly, in the latter months of the year he decided to advocate the rapid change-over to munitions production as the only alternative to the unsettled situation. There is no record of the Board's response to Dr Ferranti's renewed interest in the company, but they were certainly very much against his suggestion to manufacture shells and fuzes, and little new was tried in 1914. The Board had even decided to reduce the working week, from fifty-two hours to thirty, in mid-August and although this was rescinded in October because of pressure from the workforce, the business was clearly suffering from a lack of direction. Dr Ferranti was naturally frustrated by what he saw as the Board's apathetic attitude, but he persevered with his arguments and finally, in the early months of 1915, the Board agreed to make an initial foray into the munitions business. It was actually Dr Ferranti who secured the first contract; to machine 50,000 shrapnel shells for his former associates, Vickers. This was worth £15,000, but he was convinced that much larger contracts were available from the War Office and in February he entered into some detailed negotiations with senior civil servants to this end. It was as a result of this activity that during 1915 the Crown Works was to undergo a series of major changes. Ironically, there is great profit to be made from war, and the manufacture of shells and fuzes facilitated the company's financial recovery by providing much healthier returns than the electrical business. At the same time, it is important to understand that Dr Ferranti saw the company's transition into a munitions producer only partly as a solution to the problems in the electrical market. He was first and foremost a fervent patriot and, despite his Italian name and distinguished Italian ancestry, he regarded himself as an Englishman. In several trade papers, Dr Ferranti is sometimes depicted waving the Union Jack and clearly his contemporaries were impressed by his attitude, while other non-British industrialists, and even American-owned businesses, experienced a jingoistic backlash during the war.

In the early months of the war the government had relied on private enterprise for its supplies of ammunition, but it soon became apparent that more rigorous policies were required, particularly after the scandal of the Great Shell Shortage in the spring of 1915. This resulted in the creation of the Ministry of Munitions and the 'controlling' of over 700 engineering companies for the purposes of manufacturing larger quantities of armaments. Ferranti were controlled in September 1915 but by this time the Ministry of Munitions had already placed contracts with the company for eighteen-pounder shrapnel cases, six-inch shells and No. 100 fuzes worth £330,000. As the Crown Works was obviously not equipped with the machinery to produce these devices in such large numbers, the War Department also provided £100,125 as a mortgage on the factory to finance the purchase of machine tools and the construction of an extension measuring 220 feet by 50 feet. This involved a major change in layout and Dr Ferranti reported to his son Vincent: 'It is a very difficult business getting our work changed over onto the new lines.' It was naturally on Dr Ferranti that the burden of this work fell, not simply because he had been the only advocate of such a move, but mainly because he was the only director with any technical ability. In fact, there would appear to have been a decisive switch in the balance of power within the Board during the War, with Dr Ferranti once again taking on the mantle of prime mover. They were careful enough to use the effective monitoring device of an Executive Committee, on which sat Arthur Whittaker,

Dr Ferranti and a representative of the Ministry of Munitions, but in effect the company's founder was primarily responsible for the progress made after 1914.

The Executive Committee proved to be a flexible means of managing the company while the Crown Works was 'controlled'. Composed of an accountant, an engineer and a civil servant, it made a major contribution to Ferranti's war effort. The Ministry of Munitions had decided in 1916 to appoint its own representatives to all 'controlled' companies in order to ensure that the armaments contracts were executed efficiently and speedily. Ferranti were well served by their appointees, in the first place O. Winder between June 1916 and July 1917 and thereafter Captain P.D. Thomas. They supervised Dr Ferranti's work effectively, but without interfering in the more detailed aspects of production, and after the War Captain Thomas was invited to remain as a director, serving in this capacity until 1929. Ferranti, of course, were not completely ignorant of mass-production techniques, having been manufacturing meters along these lines since the early days at Charterhouse Square, but the production of standardised and interchangeable shells and fuzes required a tremendous attention to works organisation. In general, British industry was to benefit enormously from the experience gained during the War in working with automatic machine tools and rationalised production processes, and Ferranti were no exception. Dr Ferranti worked some long hours to perfect the necessary methods and, with his personal engineering team, led by Albert Hall and A. Tucker, he overcame the initial problems of making shells and fuzes with a consistent quality. In this way, he developed the methods of team development work which were to become such a feature of Ferranti's progress after the War. By the end of hostilities, such was the rate of progress achieved that output of six-inch shells had increased from 500 per week in September 1916 to 4,500 by December 1918, while the eighteen-pounder-shell shop saw production rise from 800 to 6,000 per week over the same period. Dr Ferranti even patented a method of hydraulically forging shells in 1918, but an indication of the effort he put into this work is his failure to take out any patents in 1916 and 1917. This was very unusual for such a prolific engineer, and only by dint of co-ordinated team work did the electrical business maintain its head above water in those difficult times.

Although the company's peace-time products were suffering both from a lack of orders and from some technical neglect, the large munitions contracts allowed Ferranti to emerge from the war in a financially stronger position. One can gauge the change in the size of its business by comparing output figures: between 1907 and 1913 Ferranti had produced goods to the value of £1·34 million but, in the years 1916–18, munitions sales alone came to £1·8 million. (An allowance for war-time inflation would only affect this comparison marginally.) On the latter, a gross profit of £347,690 was made, representing a return of almost twenty per cent, but by no means all of this was available for internal use. As we have already seen, the State had financed the expansion of the Crown Works, and by 1918 a total of £165,212 was owed to His Majesty's Government on this account. Parr's Bank had also extended the company's overdraft facilities after 1916 in order to provide the necessary working capital for the increased scale of operations. The appendix reveals that only in 1918 did the company report large net profits, of £35,174, indicating that the repayment of loans and the imposition of a bill for £54,000 as the excess profits tax on war-time profits drastically reduced the earnings. At the same time, by 1918 the Crown Works had been expanded and the company then owned a large stock of modern machine tools which were later either adapted for peace-time purposes or sold to finance new investment. Tait and Whittaker also used the profits to enlarge the reserves to £47,000 by 1918 and, in an effort to reduce the burden of interest payments, nearly £11,000 of debentures had been redeemed. This prudent policy improved the prospects for future growth, but one could never say that the company made huge profits on the munitions business. The imposition of excess profits tax and the need to repay the large loans ensured that Ferranti would never the guilty of profiteering at the taxpayers' expense. Nevertheless, Ferranti's first contact with the State as a customer brought both financial and organisational benefits to the Crown Works, furthering the recovery process started in 1905 and providing the reserves for later ventures.

The healthier signs of financial solvency in 1918 can largely be attributed to the work of the Executive Committee, and Dr Ferranti in particular. He had persuaded his fellow-directors to move into munitions production and it was largely through his efforts that the company was able to execute its contracts so successfully. This contribution is reflected in the remuneration awarded to the members of the Executive Committee in 1918 as an *ex gratia* payment: Dr Ferranti was given £1,500, while Whittaker and Thomas received only

£550 each. In 1918, Dr Ferranti was also appointed Technical Manager and head of the Executive Committee, after Whittaker had reported to Tait in a confidential memorandum that the improvement in the company's position 'was mainly due to Ferranti'. It is interesting to note that the Executive Committee was retained as a key decision-making body within the company's structure after the War, a refinement which allowed greater flexibility within the organisation. Up till 1914, Tait, Whittaker and Anderson had retained all the executive power within the Board but during and after the War the Chairman was heavily involved in other ventures. We have already noted his interest in the British Aluminium Company in the last chapter and, as Davenport-Hines (1986) points out, Tait was also part of the group responsible for the creation of the Federation of British Industries in 1916. This eminent position in industrial affairs brought him wider recognition when, in the debates about reconstruction at the end of the War, he was invited to sit on several government committees. All this activity outside Ferranti naturally reduced the amount of time he could devote to its affairs and, with Anderson dead, a power vacuum emerged which Dr Ferranti obligingly filled. Tait was initially very cautious about restoring Dr Ferranti to such a powerful position but, with Whittaker as adviser on the Executive Committee, the Chairman hoped that the new Technical Manager's well-known disposition for continual experimentation could be harnessed in a beneficial manner.

Dr Ferranti's rehabilitation was further extended in 1923, when the Board recognised his role as prime mover within the organisation by implementing a scheme of rearrangement in the company's capital structure. He had actually been badgering Tait for this change since 1921 but only in the more settled financial conditions of 1923 could £82,792 of the General Reserve be capitalised to purchase 50,000 new Ordinary Shares, 17,792 6% Preference Shares and 15,000 new 7% Non-cumulative Second Preference Shares. Other vital changes were also introduced: the existing Ordinary Shares were converted into 7% Non-cumulative Second Preference Shares, and the Deferred Shares were converted into the new Ordinary Shares, which in turn were transferred to the Deferred shareholders at the rate of five Ordinary Shares for every one Deferred Share held. In essence, this raised the issued capital from £123,374 to £206,166 and heralded the return of Dr Ferranti to a more powerful position within the company. Although he had given his eldest surviving son, Vincent, 2,000 of his

Deferred Shares in 1920, and all the Preference Shares retained the right to vote, this meant that the Ferranti family now held twenty-nine per cent of the equity. Tait still remained Executive Chairman, reflecting the Board's desire to control the company's finances along orthodox lines, but by 1923 Dr Ferranti was the majority shareholder in an enterprise which owed much to his technical genius. He was never to hold anything like the same control he had possessed prior to 1903, with the Board handling the cash-flow and departmental managers organising delivery schedules, but as Technical Manager he made the leading contribution to the company's progress after the War. It was a refinement in the ownership and structure of the business which laid the basis for a period of expansion and diversification.

Although the munitions contracts occupied so much of the company's resources, it would be misleading to say that the electrical business was completely ignored during the War. Of course, acute shortages of essential raw materials curbed production, because the Munitions of War Act, passed in July 1915, gave the State complete control of industry, and peace-time products were accorded a low priority. In 1917 Dr Ferranti complained to his son Vincent: 'Tons of copper are being needlessly thrown away and yet meters which require so little in relation to their value are being killed for the want of it.' The first part of this quotation may tell us something of his distaste for the war by this time but, regardless of his complaints to the Ministry of Munitions, electrical products were obliged to take second place behind shells and fuzes. The loss of skilled workers to the armed forces also caused problems and, by the end of 1916, unskilled female labour was being employed in the assembly of complicated switchgear. We shall discuss the company's labour relations policies in the final chapter, but again the War was causing strains on Ferranti's peace-time business and the Executive Committee was obliged to spend at least part of its time reorganising the production of electrical machinery and instruments to obviate some of these labour and raw material problems. Particular attention was paid to economising on the use of both time and materials because, despite the War, orders continued to be placed with the company. Unfortunately, no figures for electrical output are available for the years 1914–16 but, after the initial setbacks, meter orders, for example, seem to have stabilised at around forty-five per cent of their 1913 level. The agency network was actually maintained in an effective state and, in April 1916, Ferranti reported to the *Electrical Review* that orders worth £46,000 had recently been received from Europe, India and Australia. In 1917 total electrical output came to £117,000 and sixty-two per cent was exported, illustrating the value of Anderson's work in establishing the world-wide network.

Naturally, 'Business as Usual' cannot be used as a label to describe Ferranti's electrical business during the War. In fact, as a direct result of this disruption the company closed two of its departments permanently, Domestic Appliances and Switchgear. It had been difficult enough building up a trade in electric fires and cookers in peace-time, but after 1914 it proved to be an impossible marketing and production task. Adverts continued to be placed in the trade papers but by the end of 1916 the Board had decided to sell all their patents and stock relating to this business to the Jackson Electric Stove Company. Similarly, raw material shortages, increased labour costs, a dearth of orders, and the capture of the Switchgear Department Manager by the Germans while on a business trip in 1916 had made switchgear manufacture problematic. Tait had always warned that if this department should go into the red it ought to be closed, and in September 1918 Ferguson, Pailin & Co. were persuaded to take over this business. Both Ferguson and Pailin had been Ferranti employees and, like H. W. Clothier (1933), they had learnt the basics of switchgear design working with Dr Ferranti. In 1913 they had left the company to establish their own business, and their acquisition of Ferranti's trade name and patents in 1918, for £2,250, provided a major boost to their position in the switchgear market. Ferranti consequently emerged from the war with a streamlined product base of meters, instruments and transformers, but this rather misrepresents the reorganisation of the peace-time business after 1915 because by 1918 the latter had been expanded into a much larger department. It is ironic that only in the unsettled climate of a world war could Dr Ferranti begin to exploit his reputation as one of the pioneering transformer designers, but by the end of the war his company had seized the opportunity to make a major impact in this market.

Dr Ferranti, along with many other leading businessmen like the Productioneers' Movement of Dudley Docker and Sir Vincent Caillard (Davenport-Hines, 1984), was firmly of the opinion that the war should be used as a chance to eliminate German competition. He stated at a BEAMA Council meeting in 1916: 'It should be as much a part of our policy to permanently cripple our enemies in trade as it is to defeat them in war and all means to this end should be relentlessly pursued.' This is another indication of Dr Ferranti's fervent patriotism and was one of the reasons behind the company's decision to manufacture large transformers in 1915. Of course, the market stimulus was extremely important but, as Davenport-Hines (1986) has noted of Hugo Hirst's decision to produce arc-carbons and the latest light-bulbs in this period, business decisions were not always based strictly on economic criteria. In the transformer market up to 1914, the traditionally small-scale nature of the average British power station had not encouraged the development of the larger units increasingly in demand by the First World War. When this demand started to emerge after 1900, it was either the foreign conglomerates or their British subsidiaries which won all the big orders, leaving companies like Ferranti to supply the smaller end of the market. Even though Dr Ferranti had pioneered the design and construction of the world's first modern transformer in 1885, his company was only a minor producer by 1914, illustrating in the starkest manner just how unresponsive the British electrical market had been in stimulating the introduction of new technology. The war, however, not only prevented foreign manufacturers from exporting to Britain, it also stimulated the expansion of the electricity supply industry. This, in time, led to greater self-sufficiency in Britain because after the defeat of Germany a much smaller proportion of home consumption went to foreign firms, with British manufacturers having developed competitive product ranges to supply all the needs of a more sophisticated home market. This can be illustrated by quoting the statistics collected by BEAMA (1927): in 1913 British output accounted for seventy-seven and a half per cent of total sales of electrical and allied products in Britain; in the 1920s the proportion is closer to ninety per cent, with sales growing nearly three-fold from £6·8 million in 1913 to an average of £17 million between 1920 and 1927. (Some allowance for war-time inflation should be made here, but after 1920 prices of electrical goods fell by up to sixty-five per cent.) Ferranti shared in this boom but before we can discuss the transformer business we must analyse the general trends in the electricity supply industry at a crucial period in its history.

It has already been established in Chapter 2 that the Deptford system of high-tension AC electricity distribution had a limited effect on the organisation of British power station operations up to 1914.

One could argue that the strong legal powers vested in the local authorities provides the key to these developments because they were reluctant to lose their managerial prerogatives in controlling electricity supply and see private enterprise take over. This municipalisation of the industry acted as the major constraint on the industry's growth, and up to 1926 very little interregional activity was allowed. Hannah (1979) has skilfully related this story in great detail, and this section owes much to his comprehensive study of the industry. He illustrates how, even though the 1898 Joint Select Committee of the House of Lords and House of Commons, headed by Viscount Cross, had over-ruled the municipalities and accepted the need for larger power companies to distribute electricity over whole counties, the local authorities opposed any attempt to pass control of the industry into private hands. The idea of one power station per borough was technically outmoded by the turn of the century, and entrepreneurs in many parts of the country were creating power companies to exploit the full economies of scale available in a high-tension AC system. Dr Ferranti had played a minor role in this development, as we saw in the last chapter, acting as an advocate of such schemes in front of parliamentary committees. The most famous company for which he gave evidence was the Newcastle Electricity Supply Company (NESCo). Under the guidance of one of the most prominent electrical engineers of the era, Charles Merz, this enterprising venture controlled an area of over 1,400 square miles and, according to Hannah (1979), accounted for one-eighth of total electricity sales in Britain by 1914. It was an outstanding example of the way forward in electricity supply, transmitting three-phase AC at 20,000 V, but up to the First World War NESCo was an exception in Britain. Of course, LESCo was still working along the lines laid down by Dr Ferranti, but in London alone there were seventy separate power stations. Even when the power companies eventually succeeded in gaining parliamentary approval for their plans to supply in bulk over wide areas, such was the strength of municipal opposition that rarely were the newcomers given access to the large urban markets. Sales of electricity did continue to rise in this period, from 37.8 GWh in 1895 to 1976 GWh in 1913, but British electricity supply was characterised by a wide diversity of systems, operating with low load factors and controlled by the conservative municipalities. Had costs been reduced by using more efficient generating and transmitting methods, sales would undoubtedly have increased at a faster rate in a period when industry was vainly looking around for cheaper production costs. Unfortunately, it was only the development of a metal filament lamp in 1911 which brought electric lighting prices into line with gas, while electric power remained too expensive in much of the country. The scale of operations was beginning to increase after 1900, with generators of 11,000 kW capacity installed in the Carville station for example, but these only highlighted the conservatism in most parts of the industry.

It was only in the First World War that official attitudes towards the control and organisation of the industry began to change. After over thirty years of campaigning, the State was beginning to recognise the crucial importance of electricity. This is reflected in Lloyd George's creation of a Department of Electricity Power Supply, and the appointment of W. McLellan as its Director, in 1916. There was even some discussion of power station interconnections but the technical difficulties of standardising voltages and frequencies prevented immediate progress on this matter. Ballin (1946), in a detailed review of the industry's progress up to 1939, calculated that ninety-five per cent of the factories built during the war were powered by electricity, bringing an improvement in power-station loading with increased demand throughout the day. Approximately £23 million was spent on electricity supply as sales jumped by eighty per cent between 1913 and 1918 to 3,570 GWh and, in the debate about reconstruction, there was general recognition that electricity would play a crucial role in Britain's economic future. Charles Merz was a great influence on official opinion concerning this issue but he warned the Cabinet's Reconstruction Committee that the institutional barriers to interconnecting power stations must first be removed. This seemed to indicate a trend towards greater integration of the industry and even the powerful local authority grouping, the Municipal Electrical Association, pledged its support for reform. Plans were drawn up for a Bill in 1919, and Christopher Addison (the Minister for Reconstruction) stated in its first reading to the House of Commons: 'An adequate and cheap power supply, widely distributed throughout the country, will open up possibilities comparable with those of the industrial revolution a century ago.' At last politicians were moving closer to the views expressed by men like Merz, McLellan and Ferranti, but, in the post-war government's haste to dismantle all the paraphernalia of State control, the 1919 Bill resulted in an unsatisfactory compromise. The aim of the original Bill had been to create District Electricity Boards run by Electricity Commissioners with compulsory powers to coerce competing interests to merge, but the right-wing tore the heart out of the proposals by granting little authority to the new bodies. This simply perpetuated all the old problems because the municipalities forgot all their war-time interest in interconnections and no integration took place under the voluntary arrangements. It was an opportunity sadly missed for the politicians to bring the industry up to date.

The political failure of 1919 can only be understood with reference to the wave of anti-interventionism which spread throughout political and business circles after the virtual nationalisation of the economy in the war. There were also some who still argued that, in a country blessed with abundant supplies of coal, long-distance transmission lines were unnecesary, but such opinions went against the trend in the world generally, with the establishment of highly co-ordinated schemes in many countries. Canada and New Zealand were two which had recently invested heavily in large-scale networks, employing the full advantages of high-tension AC systems. In the USA and Europe, extensive use of this technology was also increasing, while British supply engineers in general held back. In his presidential address to the IEE in 1910, Dr Ferranti had put forward a radical new programme for the 'All Electric' future, with 100 super-power stations each with a capacity of 25 MW, located close to cheap sources of fuel on the coalfields, and a national distribution network capable of supplying the needs of both domestic and industrial consumers. He also hoped that more effective use of coal by-products would be part of this programme, as a means of stimulating activity in other industries, but his principal concern was with making electricity the fuel of the future at a price that would benefit the British economy. Many contemporaries ridiculed the economics behind these plans — at that time the average size of generators in Britain was under 1 MW (1 MW equals one million watts) — and unfair comparisons were made with the commercial fortunes of his Deptford station but, in the context of Merz's achievement with NESCo and the progress made using high-tension voltages in other countries, Dr Ferranti's 1910 scheme was by then more commercially viable. Sir Leo Chiozza Money, the prominent economist, claimed in 1919 that such a scheme 'would be fruitful almost beyond imagining' and Hannah (1979) has concluded that Dr Ferranti's speech was 'a tribute to the value of scientific imagination' because 'much of his futuristic vision was to materialise'. Unfortunately, the 1919 legislation failed to provide the statutory powers for such a reorganisation, but in the 1920s greater progress was to be made, culminating in the establishment of a scheme similar to that proposed by Dr Ferranti in 1910.

After 1919, the outmoded organisation of the British electricity supply industry became even more apparent, particularly as many influential contemporaries were advocating the greater use of electrification to improve industrial competitiveness in the severe depression of the early 1920s. Consumption of electricity actually continued to increase, from 3,570 GWh in 1918 to 5,817 GWh in 1926, but the Electricity Commissioners were powerless to implement the much-vaunted policies of rationalisation and integration. There was still a multifarious variety of systems in use and, according to the returns of the Electricity Commissioners in 1926, of the 597 stations under their jurisdiction in Great Britain, fifty-six per cent were only capable of generating a maximum of 2·5 million kWh, while the extremely low number of only fifteen could generate more than 100 million kWh. This situation was regarded by most leading industrialists, and belatedly by politicians, as completely unsatisfactory because low capacity stations could not provide cheaper electricity. Finally, in 1924, Lord Weir was appointed by Stanley Baldwin, the Conservative Prime Minister, to review the industry's progress and suggest possible remedies for any faults. Lord Weir, a prominent marine engine components manufacturer, was one of those who recognised the need for cheap power to improve the competitiveness of British industry – electricity in Britain was twenty-four per cent more expensive than in the USA, and up to four times the price of Swiss, Italian and French electricity – and he warned that fundamental changes costing approximately £35 million would be required if the industry was to be brought up to the standards set abroad. Predictably, the local authorities objected to his report of May 1925, largely on the grounds that Weir's proposal to create a Central Electricity Board would significantly reduce their traditional powers. There were also loud complaints about excessive State intervention, but in 1925 a stronger political will existed and consequently the 1926 Electricity (Supply) Act was passed. This was a far cry from nationalising the industry because 'selected' power stations, both privately and local-authority-owned, would supply electricity for transmission around a national grid financed by the State-sponsored Central Electricity Board (CEB). Nevertheless, it was a radical step to overcome the problems of the past and in the next fifteen years the CEB had a major impact on the electricity supply and electrical engineering industries in Britain.

The 1926 Act was undoubtedly a significant milestone in the history of British electricity supply, and many forces had combined to influence its formulation and passing into law. The Conservative cabinet of 1925, Sir Eric Geddes, as Minister responsible for the Bill, and Lord Weir were the leading figures in the events of those years, while on the industry's side Charles Merz, W. McLellan and Sir John Kennedy stand out as the important influences on policy. In this context, Dr Ferranti can be regarded as one of the earliest voices pushing for this change. In the widely-reported adventures at the Grosvenor Gallery and Deptford, in the much-applauded speech of 1910, and in his industrious support of movements like the British Electrical Development Association, which we shall discuss in the next chapter, he acted as one of the strongest influences on the eventual victory of the high-tension AC system. The CEB's grid was also close to the ideas Dr Ferranti had postulated in 1910, and although 144 power stations were 'selected', as opposed to his notional figure of 100, it signifies the validity of his vision for the future of electricity supply. He clearly had no direct influence on Lord Weir's report, because at no time was he consulted, but Dr Ferranti's role in creating the climate for these changes was recognised by his contemporaries as of seminal importance. There are constant allusions in the trade papers of the 1920s to his pioneering venture at Deptford, and it was fitting that his firm was to play a leading part in designing and producing a vital component of the grid system. This was to be a source of great satisfaction to him in the last years of his life because the 'All Electric' ideal seemed to be closer at hand.

One of the more interesting features of the grid network, which the CEB was to build between 1929 and 1932, was the decision to use 132,000 V as the transmission voltage. This was unprecedented in Britain and called into question the ability of manufacturers to produce equipment of the required sophistication. (It was generally accepted that only British-based companies would be given the business.) One of the crucial aspects of the 132,000 V system, apart from the cable, would be the transformer, and few British manufacturers prior to 1914 would have been equipped with the technology to handle such a pressure. During the war, however, Ferranti, along with other specialist manufacturers, had begun the development of more advanced designs. The growing use of electricity, especially in industry, and the provision of a guaranteed home market stimulated more ambitious efforts in this field, and Dr Ferranti was amongst the most innovative. The production of transformers had actually been

a sideline of the Instrument Department since the move to Hollinwood in 1896 and, even though Dr Ferranti had pioneered their design
in the 1880s, the business remained small-scale. The largest Ferranti
transformer manufactured had been for Kilmarnock, with a capacity
of 800 kVA (kilo-volt-amperes), but, after a development programme
had been started in 1915, orders were being taken from Manchester
Corporation for 1,000 kVA units and in 1918 Birmingham Corporation ordered seven with a capacity of 1,250 kVA. This expertise was
putting Ferranti at least on a par with the general electrical manufacturers like Metro-Vick's and BTH and the specialist transformer
manufacturers like Berry and Fuller in what was an expanding
business after 1915.

It was Dr Ferranti again who was responsible for this strategic
switch in direction by the company. He recognised the opening in
the market and he designed the larger units demanded by British
customers. Typically, he was anxious to improve the mechanical
construction of the transformer, patenting in 1915 and 1918 a series
of improvements to the casing, the springs holding the induction coils,
and the method of cooling. New methods of impregnating the insulation were also introduced to avoid the accumulation of damaging
moisture and, by the end of the war, new testing equipment was being
designed to guarantee the safety of the insulation. Such was the company's reputation in the market by 1918 that the Board was willing
to sponsor Dr Ferranti's plan, as Technical Manager, to establish
a High Voltage Laboratory in the Crown Works, where further
development work could continue the progress already made. This
laboratory was staffed with a design team which included one of the
leading British transformer experts, E. T. Norris, and in 1919 Ferranti's commitment to transformer technology was further illustrated
by its acquisition of the important Lulops patent for oil-immersed,
water-cooled transformers. Dr Ferranti by this time was not averse
to enlisting outside assistance for his projects, as we saw on the
turbine, and as a result the High Voltage Laboratory became an
important centre of excellence in transformer design. It also reflects
the new approach to product development emerging at Ferranti during
this period, with the Technical Manager supervising a series of team
projects in tailor-made facilities. The company had always pursued
extensive product development, but after 1915, not only were the
funds available but the management structure had been adapted to
facilitate this work. We shall see later how this approach to team

development work was transposed onto Ferranti's ventures in the 1920s, but in this period it provided the expertise for a successful attack on an expanding market for large transformers.

The investment in improved transformer development facilities was to prove of lasting benefit to the company after the First World War. Output of transformers had already exceeded £155,000 in 1919, compared to £45,000 in 1917, and the Board agreed with Dr Ferranti's recommendation that a Transformer Department should be created by investing £10,000 in new plant. This was financed by the sale of surplus plant from the redundant munitions shops, which realised £12,000, but by 1920 further extensions were required as transformer output increased to over £230,000. By this time, the High Voltage Laboratory was gaining a wide reputation for its experimental work and in 1919 the Berry Electric Transformer Company had approached Ferranti with an offer to merge their interests. This, as we shall see later, was all part of the concentration movement affecting the electrical industry after the War, but not even the leading company promoter, F. A. Szarvasy (of the British, Foreign and Colonial Corporation), could convince Dr Ferranti and his Board that there were advantages to be gained from a merger. They preferred to remain independent and in the 1920s continued to investigate the possibilities of further increasing the size of transformers as British and foreign customers started placing orders for even larger units. Again, the company did not pursue an introverted policy and in March 1924 Dr Ferranti travelled to Switzerland to purchase the right to use the Dessauer patent owned by the Haefely Company covering the connection of transformers in 'cascade'. This form of testing transformer involved several transformers, graduated in size, to increase the pressure on the insulation, and by 1925 Ferranti had constructed a one-million-volt unit in the Crown Works which produced the first one-million-volt arc ever seen in Britain. The *Electrical Review* of March 1925 noted that Ferranti's achievement was 'of great historical interest', considering Dr Ferranti's work in the high voltage field over forty years previously, and it served to establish the company as the leading manufacturer of testing transformers in Britain. During the 1920s several prestigious contracts were placed with Ferranti for these units, including the three one-million-volt machines ordered by GEC, Steatite & Porcelain and Callenders' Cables, reinforcing the company's growing reputation in the transformer market generally.

The purchase of such large testing transformers by other electrical and allied manufacturers in the 1920s is a clear indication of another important trend in which Ferranti participated, namely, the growing sophistication of research facilities in British industry. Sanderson (1972) has commented that in the inter-war period, despite the depressed condition of some parts of the economy, 'British industry ... experienced rapid scientific development', and the electrical industry benefited more than most from this attention to future product design. One of the more important pioneers in this respect was British-Westinghouse, who established a Research Department in 1917 under A. P. M. Fleming, and from that date the company (later to become the Metropolitan—Vickers Electrical Company in 1919) made a leading contribution to the furtherance of scientific and technical training in Britain. Other examples of the industry's greater interest in research include the large laboratories built by GEC at Wembley and Siemens' facilities in Preston, illustrating how, as Sanderson (1972) argues, the source of new developments in electrical technology was moving away from the universities into well-organised research departments staffed by trained technicians and science graduates. The State had also recognised during the war that research should be encouraged, establishing the Department of Scientific and Industrial Research in 1915 to provide assistance. Of course, Dr Ferranti had always seen this philosophy as an essential feature of his company's strategy, perhaps to too great an extent at times, and in the development of larger transformers he had provided Ferranti with a reputation second to none in Britain, bringing significant benefits when the big CEB contracts were awarded in the late 1920s.

Dr Ferranti's role in the Transformer Department's progress was obviously of vital importance, but the Ferranti family contribution did not stop there because by 1921 his oldest surviving son, Vincent, had been appointed manager of this new venture. Tragically, in July 1917, the first Ferranti son, Basil, had been killed by a direct hit on his tent in the early hours of the morning, leaving Vincent to take over as heir-apparent. Both Basil and Vincent had seen extensive action since joining the army in 1914 and each had been awarded the Military Cross, but the former's death naturally came as a great shock to the Ferranti family. Gertrude (1932) remembers how: 'It was the first terribly sad thing that had happened to us in our married life.' Although Basil had never played an active role at the Crown Works, he had received a thorough engineering apprenticeship and was seen as Dr Ferranti's natural successor. Now, as Dr Ferranti informed

Vincent: 'You ... will have all the more to do to help me, and a hard time when you get back, but I am sure you will succeed alright.' As we saw in the last chapter, Dr Ferranti had been careful to instill into his two eldest sons a liking for technology, providing both with a sound education at Repton and two years of training at Yarrow's shipyards at Scotstoun. Clearly, no second generation malaise can be discerned in this family. Vincent returned from his military service prepared to commit himself wholeheartedly to helping his father build up a flourishing business, purchasing 1,800 Ordinary Shares in Ferranti as an indication of his sincerity. Francis Ince warned him that such an investment would never make his fortune, but Vincent typically replied that he now had an incentive to help put the business on a sound footing. One must remember that Vincent's formative years were dominated by his father's exile from the company, and he was always anxious that this should never happen to the family again. This illustrates just how different Vincent's approach to business affairs was to prove, compared to Dr Ferranti's, in that, while both recognised the value of engineering excellence, the former was much more tuned in to the cash-flow problems of a family firm. Of course, it was only after Dr Ferranti's death that such a policy could emerge, and in 1921 Tait was sceptical of Vincent's ability, at the age of 27, to manage an expanding and technically sophisticated business like the Transformer Department. After leading a gunnery platoon for almost two years, Vincent, on the other hand, was sure that he had the necessary organisational and leadership qualities, and his business career illustrates just how successful Dr Ferranti had been in providing for his succession (Wilson, 1984).

The Ferranti Company had found a first-rate manager for their new Transformer Department, because it was largely through his work, in association with the technical contributions of Dr Ferranti and E. T. Norris, that the company made such rapid strides over the next decade. Just to illustrate that he was not simply an organiser, we should also note how in the 1920s Vincent and E. T. Norris jointly developed more effective surge absorbers capable of reducing the stress on transformers caused by lightning strikes or short-circuits by up to eighty-five per cent. This provided another important form of protection and, when combined with the High Voltage Laboratory's work on testing transformers, it complemented Ferranti's ability to produce some of the largest units seen in Britain to date. One of the most notable contracts of the early 1920s was for seven 4,000 kVA transformers for a New Zealand customer, to be used in stepping up the voltage from 11,000 V to what as an unprecedented level for a British manufacturer of 110,000 V. The *Electrical Review* of October 1922 commented on Ferranti's ability to produce 'equipment that, as yet, there was no demand for at home', refuting any past criticisms that British companies were incapable of manufacturing such advanced products. Further significant orders included the sale of twelve 10,000 kVA transformers to the Tata Power Company in India, while between 1923 and 1930 the Victoria Falls & Transvaal Power Company of South Africa ordered transformers worth £105,000, ranging in size between 500 and 20,000 kVA. An added reason why the latter placed so much business with Ferranti was their Chief Engineer's former position as a member of Dr Ferranti's experimental team up to 1903, but sentiment would have played only a small part in the decision to rely on Ferranti's technical expertise.

The world-wide reputation enjoyed by Ferranti transformers was also shared by the Canadian subsidiary in the 1920s. Since its establishment in 1912, the company had been steadily building up goodwill in its home market, and the restrictions on shipping and exports from Britain during the War forced them to start manufacturing their own products. Using Ferranti designs, but adapting them to the technical requirements of the Canadian system, the business expanded to such an extent that larger premises had been taken in Toronto by 1919. To reflect its new function, the agency was renamed the Ferranti Meter & Transformer Manufacturing Company, and by 1920 orders worth $140,000 had been won for transformers alone. Although the Canadian company never made large profits in this period – the profits for the whole of the 1920s did not exceed $35,000 – some significant successes were achieved in the transformer market: in 1922 the Toronto Hydro-Electric Commission ordered 100 25 kVA units and in the same year they won the largest transformer export contract recorded in Canada to date, for 200 11,000 V units for New Zealand. A purpose-built factory was later provided in 1930 to cope with the ever-increasing orders, and Ferranti Canada, as it became known in 1926, stands out as one of the most successful ventures into foreign markets by the company. On top of the other prestigious export contracts secured by Ferranti in the 1920s, the subsidiary's achievements added further to the company's strong position in this market.

By the time the CEB was discussing its plans for the national grid in 1927, Ferranti were ideally placed to exploit the decision to use

132,000 V as the transmission voltage. They were, after all, the only British manufacturer to have produced transformers anywhere near that voltage, and Dr Ferranti and Vincent were influential in determining the type of unit the Electricity Commissioners eventually chose. Initially it had been decided that single-phase transformers should be used on the grid, but Dr Ferranti and his Transformer Department manager visited the CEB in 1927 to persuade the Electricity Commissioners that three-phase transformers would produce greater economies in terms of transmission costs. This would mean that the CEB would require only one transformer, rather than a combination of three single-phase transformers for the three-phase AC used on the grid. As the grid was established largely as a means of improving the competitiveness of British industry by providing cheaper electricity, their suggestion made great sense, and it was finally decided that three-phase transformers should be adopted in place of the less efficient single-phase units. This decision, and the Ferranti role in its determination, upset most of the other transformer manufacturers because they had all agreed to quote collectively only for single-phase units. In fact, Ferranti had decided in September 1922 to withdraw from the transformer price ring, largely on the grounds that the restrictive practices were putting off customers. This move seems to contradict what was said in the last chapter about the company's role in the formation of BEAMA and, although the meter price arrangements were maintained effectively, Ferranti do seem to have acted in a disruptive manner in the transformer market of the 1920s. BEAMA (1927) has shown how the price depression of the 1920s, a trend we shall examine later, slowly broke down the industry's grip on the market, and by 1927 most of the price rings had been temporarily disbanded. Even the 1926 Act could not bring a return to solidarity, such was the mistrust among manufacturers, and in this context Dr Ferranti's stand on three-phase transformers does not seem so unusual. It also illustrates his willingness to put the cause of cheaper electricity before the interests of the industry. Above all, he wanted an 'All Electric' future, and in this instance his firm was to benefit from his independent stand.

The combination of the Ferranti role in the decision to use three-phase transformers and the company's wide reputation as manufacturers of the most sophisticated units in Britain provided them with a natural entrée into the CEB market. The construction of the grid did not actually begin until 1929 and the transformer contracts were only placed in the middle of that year, but according to Ferranti's own estimates they were awarded thirty-two per cent of the £1·5 million worth of business in 132,000 V and 66,000 V transformers by the end of 1930. Their nearest rivals, Berry, received approximately twenty-seven per cent of the orders, while companies like BTH could only claim fourteen per cent and Fuller eight per cent. In total, between 1929 and 1932, the value of Ferranti's transformer orders from the CEB and 'selected' stations exceeded £780,000, putting them well ahead of their rivals for this business. The company also manufactured the largest transformers ever seen in Britain at that time, with a rating of 80,000 kVA, and Vincent's team executed these contracts successfully and delivered the units on time. It was a technical and commercial success for which Ferranti had worked since 1915, and Dr Ferranti's role in the design and construction of a system he had advocated for many years was a fitting end to a career dominated by this desire to see the 'All Electric' ideal make real progress. After over forty years of missionary work, at last a grid was being constructed and even the State was providing assistance in extending the use of electricity to a larger number of industrial and domestic consumers. Dr Ferranti was to die before the grid actually came into operation, but in the 1920s, just as in the 1880s, he had been working actively to encourage more people to use electricity, and his efforts had not been in vain. The success in designing large transformers also reflects the true value of his technical genius, and the establishment of the Transformer Department can be regarded as an outstanding achievement. We shall examine the financial implications of this strategy later, but in boosting Ferranti's reputation around the world it confirmed Dr Ferranti's position as one of the most important British contributors to the development of power station systems.

The construction of the grid network, and the placing of all the contracts with British-based companies, was a major fillip to the British electrical industry, coming as it did just as the economy was slipping into the major depression of the years 1929–32. Catterall (1979) has shown how even in the worst year of 1931, when all British industries had experienced a 6·4 per cent drop in output, only a 3·9 per cent fall in electrical output was recorded. When one considers that total British electrical exports had fallen by almost twice that rate, the vital role played by the CEB is further substantiated. The electrical industry had actually grown considerably since the pre-War years: in 1907, according to Catterall (1979), total electrical output stood at almost £13·9 million, but by 1930 this had leapt to over £87·7 million. There had been some severe vicissitudes, particularly during the War, but more alarmingly in the early 1920s when British consumption of electrical machinery fell from £19 million in 1921 to only £10·7 million in 1922 (BEAMA, 1927). This, as we have already noted, precipitated ruinous price competition which was not helped by the collapse in general export prices, from 303 in 1920 to 188 by 1929 (1913 = 100). The appendix reveals how Ferranti output fared in this period, despite the calamitous price fall of the early 1920s, rising from £207,704 in 1918 to over £1·2 million in the year to June 1930. The agency network also ensured that exports would continue to increase over these years, from £78,000 to almost £250,000 respectively, illustrating how Ferranti continued to share in the boost to demand experienced after the War. At least seventy-five per cent of all the company's foreign business was actually done in Empire countries in the 1920s, re-emphasising the point made in the last chapter about its dependence on these markets. Nevertheless, with the expansion in electricity supply all over the world, more sophisticated products were being purchased by the Empire countries, as we have seen with the transformer business, and Ferranti benefited enormously from their contacts with these countries throughout the inter-war era. It was a period of opportunity for the electrical manufacturer and it was essential to keep a close contact with the trends in the market.

The British electrical industry was clearly expanding at a more rapid rate than at any other time in its history by the 1920s, but it is difficult to generalise about a single industry in the inter-war years. One of the trends associated with the growth in output was the extensive diversification of product range, and Ferranti can be regarded as a microcosm of what was happening generally. While the transformer business was expanded, and the Meter Department continued to benefit from the attentions of Dr Ferranti and his team, Ferranti also diversified into one of the growth sectors of the inter-war period, the manufacture of radio equipment. This represented a considerable risk for a medium-sized company like Ferranti, but it was to have a major impact on the orientation of the business by the 1930s. Originating with the simple audio-frequency transformer in 1924, within fifteen years Ferranti had established a wide-ranging electronics business including radios, television, valves, cathode ray tubes and specialised devices. It was a move which converted the company from a traditional electrical engineering business into a broadly-based manufacturer, at a time when all the leading companies were following a similar path. It reflected the importance of planned development programmes in the formulation of company strategy, a policy Dr Ferranti had by now imposed on the Board after his success with large transformers.

The emergence of the radio industry was one of the most important industrial and social phenomena of the inter-war years, influencing the activities of both manufacturers and consumers to a significant extent. After the creation of the British Broadcasting Company in 1922 (later changed to British Broadcasting Corporation in 1925), demand for this new form of entertainment increased rapidly, and in 1930 over 3·4 million households had purchased licences to receive broadcasts. Dr Ferranti was characteristically fascinated by these new devices, and in the early months of 1923 he first took home a radio in order to dissect all its working parts. It would have been extremely expensive for the company to venture into the production of radios in the early 1920s, partly because of the cost of building up a business from scratch but mainly because of the nature of the radio industry. In the first place, the manufacture of radio sets in Britain was dominated by the Marconi Company up to the early 1930s, while the predominance of radio 'hams' in the market made it essential that extensive goodwill with a large number of customers would have to be earned. Apart from its founder's role in developing the first effective means of transmitting and receiving messages using only electromagnetic waves by 1896, the Marconi Company had acquired the rights to thirteen master patents covering the major components in a radio. These patents were incorporated into a licence which other manufacturers could purchase on payment of 12*s* 6*d* per valve-holder manufactured as a royalty. In the 1920s, as Sturmey (1958) argues, this severely restricted the number of companies willing to produce complete radios, although the home constructors' market remained much more important up to the end of the decade anyway. These factors strongly influenced Dr Ferranti's decision to concentrate on the components market when, after bringing his extensive experience to bear on the design of induction coils, he introduced a new type of audio-frequency transformer in 1923.

In the radio, an audio-frequency (AF) transformer is required to give equal amplification over a range between sixteen and 5,000 cycles, but even Marconi's 'Ideal' product could only produce 250 to 4,000 cycles. Dr Ferranti, using the Crown Works facilities at his disposal as Technical Manager, asked his development team leader, Albert Hall, to conduct some experiments on amplification curves and, on the basis of these calculations, a more effective design was conceptualised. Hall advised that 30,000 turns of wire would have to be placed around a tiny coil capable of inclusion in a radio set if the required levels of amplification were to be achieved, but Dr Ferranti was not overawed by the mechanical difficulties this presented. His first task was to design thermoplastically-moulded coils and, by adapting the small air turbine used in the textile machinery experiments, he devised an efficient method of automatically winding on these 30,000 turns. By the end of 1923, the company's first radio component, the AF2, was ready for the market, and sales started to increase slowly to around £1,500 per month. This was only a minor success in the highly competitive components market, but within eighteen months the AF2 had been replaced by the highly successful AF3, one of the most effective transformers of its kind with an equal amplification over a range between fifty and 8,000 cycles. It was advertised as the 'Nearly Perfect Transformer' – Dr Ferranti refused to allow the use of the word 'perfect' in publicity material – and made an immediate impact on home constructors. It was also sold at 25*s*, undercutting the Marconi 'Ideal' by almost one-third, such was the efficiency of Dr Ferranti's production methods, and by 1925 over 10,000 per month were being produced to satisfy the growing demand. Ferranti recognised in the home constructors' market a potentially lucrative business and, building on the goodwill created with the AF3, by 1927 the range had been extended to include loud-speaker units and condensers. A regular pamphlet, *True Radio Reproduction*, was even published, giving the 'ham' a step-by-step guide to radio and helping to extend the company's reputation in a period of rapid growth.

The success of this new venture placed a great strain on the Crown Works, with the output of radio components topping £100,000 in 1926. The Meter Department had been assigned the task of producing the AF3 but, with the expansion anticipated in the company's traditional products following the creation of the CEB, it was thought advisable that alternative arrangements be made for the newly-created Radio Department. The firm could not afford to build an extension to the Crown Works and, instead, by October 1926 a lease had been taken on the Eagle Iron Works in nearby Stalybridge. This was to become the home of the Radio Department for the next eight years, and over £25,000 was spent equipping these works with plant and machinery. The most interesting feature of these new facilities was what the *Electrical Review* of September 1927 described as 'one of the finest laboratories solely devoted to radio work in this country'. The Radio Department also marks a new approach by Ferranti to their research and development in the acquisition as consultants of Dr L. S. Palmer, Head of Pure and Applied Physics at Manchester University, and Professor G. I. Finch of the Chemical Technology Department of Imperial College, London. In the past, Ferranti had always hired specialists as direct employees, but they recognised the need to catch up with the major companies by hiring the services of these consultants to supplement their own development work. This was absolutely essential if Ferranti were to keep up with the rapid rate of technical progress in the design of radio components and circuitry in the 1920s, and Dr Ferranti's role as Technical Manager was to ensure that the company's product range remained viable. It is clear that he had now established continuous and planned product development as an intrinsic feature of Ferranti's overall strategy by this time and, as with the transformer, this was to lead them into new areas in the decades ahead.

The investment in the Stalybridge factory soon proved rewarding, allowing output to increase from the £100,000 of 1926 to £197,000 in the year to June 1930. Sales of the AF3, significantly, accounted for almost sixty per cent of turnover, and the trading profit on this business remained steady, averaging around £25,000 per annum, as more effort was spent on new products and designs. The radio engineers were particularly interested in designing a complete Ferranti radio and as early as September 1926 Dr Ferranti had written in his notes: 'The thing most wanted is a complete set.' He even approached the Marconi Company in 1928 to negotiate terms on a licence, but it was only in 1929 that the burdensome licensing system came under pressure after a court-case involving the Brownie Wireless Company. Although the Marconi Company won the case and Brownie were obliged to pay the full royalty for all thirteen patents covered in the licence, they were magnanimous in victory and decided to reduce the royalty from 12*s* 6*d* to 5*s* per valve-holder, on condition that companies would sign a five-year contract. The reason behind this move was that by 1934 all the patents would have expired, but at least it created a more favourable environment for other radio manufacturers. The rapid progress in circuit design had also put the construction of efficient receivers beyond the growing number of licence-holders in the late 1920s, providing an added inducement to companies like Ferranti to produce a complete radio. Having established goodwill in the market with their components, in the autumn of 1929 Ferranti consequently launched their first radio at the annual Olympia Exhibition. (Olympia provided the opportunity for manufacturers to introduce their new products before the busiest season in the winter.) This model, marketed as the Standard Model 21 and priced at £16, had been developed by a team led by Albert Hall, bringing Ferranti into competition with all the specialist manufacturers that had recently sprung up like Marconi, Bush, Decca, Pye and Cossor. It was an even more competitive market than that for components, but Ferranti hoped to enhance its position by continued development work and close attention to sensitivity.

The first venture into set manufacture was not spectacularly successful, with less than 1,000 being sold, but in 1929 more plans were being formulated by Vincent to improve the performance of the Radio Department by adding valve production to its range of capabilities. This was a crucially important decision, because the valve (or thermionic emission valve, to give it its full name) had become the fundamental building brick in circuit design by the 1920s. Its origins lay in the work of a British scientist, J. A. Fleming, in 1904 on the two-element amplifying valve, and the production of the more effective triode valve by the American industrialist, Lee de Forest, in 1907. By the end of the First World War, as Fleming (1930) recounts, it had been developed to the stage whereby most radios were then using such devices in large numbers to improve sensitivity, and some of the large multinational bulb manufacturers like Western Electric and Philipps, as well as specialists like Marconi-Osram, were investing heavily in new production plant. This made entry into the valve market expensive if a Ferranti valve was to be produced from scratch, but here the assistance of Professor Finch proved invaluable. At the Physical Society's Exhibition in January 1929, Professor Finch and two of his staff, R. W. Sutton and M. E. Sions, displayed the results of their investigations into the creation of a vacuum in valves. Vincent attended this exhibition and within weeks he had negotiated with Professor Finch to bring Sutton to Stalybridge as manager of a new Valve Department, with Sions as his research engineer. Another of Finch's former students, A. L. Chilcot, was recruited from GEC and, with the addition of Ferranti engineers like M. K. Taylor, a development team was assembled capable of producing a wide range of valves. In fact, it was to take nearly two years before full production could commence, but this team provided the nucleus of a venture which took the company into uncharted territory as far as British technology was concerned, putting Ferranti at the heart of some of the most advanced electronics projects in the decades to come.

The Radio Department had brought a new dimension to the activities of Ferranti and, although the work of the Radio Laboratory and the Valve Department only reflected the company's approach to product development, the marketing of consumer goods was completely alien to the traditional practice of selling to electricity supply authorities. In fact, such was the strain that this imposed on the Sales Department that major changes were introduced to complement Dr Ferranti's technology-led strategy. A.B. Cooper, the General Manager of Ferranti Canada, was highly critical of what he saw as an extremely passive approach in the company's marketing. With the exception of the AF3, Ferranti were regarded as slow to introduce products on to the market, largely because there was little communication between the Sales Department and the development teams. They were not averse to spending on advertising and catalogues, with an average of £13,000 per annum going on this budget in the late 1920s, but more could have been done to improve co-ordination within the organisation. This was particularly true after the introduction of complete sets in 1929, and to facilitate progress in this area a Radio Conference was inaugurated to bring together the department's designers, engineers and salesmen. More servicemen were also hired and cars were provided for the sales staff in order to extend the area covered in the home market, bringing a new sense of urgency to the work of the Sales Department. It was an aggressive marketing strategy based on the need to make more intimate contact with the consumer, and it is a reflection of Ferranti's commitment to this new venture that so much effort was put into diversifying away from the traditional product base.

Although Dr Ferranti and Vincent were anxious to push Ferranti into this new growth sector and in 1934 £130,000 was spent on a new factory at Moston, the Radio Department was never a great commercial success after the move into complete sets. Even the acquisition of a licence to produce the highly-sensitive superheterodyne circuit from the major American company Hazeltine in 1931 could not increase the annual sales of radios above 65,000, at a time when total British output was pushing the two million mark. The development of Ferranti-designed valves also failed to boost sales but, to re-emphasise an earlier point, the venture into radio was more important as a means of introducing the company to the potential in electronics. Apart from the work in the Radio and Valve Laboratories, Ferranti sponsored a three-year project to examine the possibilities in a system of ultra-short-wave wireless telegraphy in 1928. They spent around £10,000 on these experiments and Dr Palmer worked continuously with the development team, but little of practical benefit came of the work in the short term. On the other hand, by 1938 the War Office had commissioned Ferranti to work with the Department of Scientific and Industrial Research on the pioneering experiments conducted by Sir Robert Watson-Watt on radar. The company's known expertise in ultra-short-wave wireless telegraphy, and its emphasis on research into new applications for electronics, had provided an entrée into another of the most important sectors of the industry. It is often said that research work is risky because there is no certainty about the outcome, but Ferranti's policy, based on Dr Ferranti's commitment to product development, was to examine the possible applications of new ideas and inventions in the hope that innovation might follow. We have seen how successful this had been with large transformers, and with radio, although the main business was a commercial failure, it provided untold opportunities in the years ahead. In 1936, Professor Finch published a book, *Researches in Combustion Catalysis and Structure of Surfaces*, which revealed Ferranti's interest in such areas as cathode ray tubes and specialised valves. By that time, Ferranti were venturing into new fields like radar which were only just being opened up by scientists and engineers around the world, and over the next two generations this was to become the company's stock-in-trade. Other electrical companies were, of course, pursuing similar projects, and the larger resources of firms like GEC, English Electric and AEI were providing even greater opportunities, but Ferranti were by no means to be left behind in the electronics revolution of the next fifty years.

The story of Ferranti's electronics work is taking us away from the main subject of this book, but it is important in illustrating Dr Ferranti's contribution to the technology-led strategy which became such an intrinsic feature of his company's philosophy. Ever since his early arguments with Francis Ince in 1883 over the need to complete alternators for delivery without allowing for continuous improvements, Dr Ferranti had been responsible for making product development, and the search for new products, central to what became a company catch-phrase, the 'Ferranti Spirit'. By the 1920s, with the establishment of a post of Technical Manager, the creation of specialised research departments, the recruitment of a highly selected body of engineers, and the use of academic scientists as consultants, this philosophy had clearly impressed itself on the structure and outlook of the company. Yet, with the inclusion of accountants like Tait, Whittaker and Cooper on the Board, this technology-led strategy was now complemented by both an extensive agency network and a desire for financial solvency, assets which had been lacking in the past. Balance became a key word in the decision-making process at Ferranti, with directors representing a variety of different interests, but one aim above all became apparent; the desire to venture into new areas of electrical technology. This was the legacy of Dr Ferranti to his family, and after 1930 Vincent was keen to retain this combination of technical and business aims in his period as Chairman (1930–63).

The period since 1914 had been littered with opportunities for electrical manufacturers and Dr Ferranti, operating within the limitations imposed by the size and financial resources of his company, had taken as much advantage of the changing environment as most of his contemporaries. An important point to bear in mind with regard to Ferranti's progress in this period was that internal resources financed the expansion and diversification. In contrast to many of the leading companies in the industry, and in spite of the unwanted attentions of the Berry Company in 1919, Ferranti was not involved in any of the mergers of the inter-war era which created the three big companies, GEC, English Electric and AEI. It was immediately after the war that electrical engineering became involved in this merger movement, with the large defence contractors eagerly searching for new markets in which to invest the earnings from their successful munitions and armaments contracts. This resulted in Cammell Laird leading a group which combined Dick Kerr and Siemens into the English Electric Company in 1919, while in 1918, after the Americans had sold British-Westinghouse to Dudley Docker a year earlier, Vickers acquired Docker's interests to create the Metropolitan-Vickers Electrical Company (Davenport-Hines, 1984). These mergers were symptomatic of the growing scale of business prevailing in British industry by the

1920s, but later in the decade greater changes were to be made, largely through the intervention of American interests once again (Jones & Marriot, 1972). In fact, because they were poorly managed, neither English Electric nor Metropolitan-Vickers (Metro-Vicks) were very successful and, by the late 1920s, financial support was being canvassed to keep the businesses alive. English Electric, in particular, experienced a traumatic first decade and it was only after Westinghouse stepped in in 1929 that the organisation stabilised its financial position. It was at this time that General Electric was considering the acquisition of the other leading British electrical companies and, although Hugo Hirst managed to keep GEC independent of foreign control, by 1928 Metro-Vicks and BTH had been merged into the newly-created Associated Electrical Industries (AEI). By 1930, despite this American intervention, GEC had actually become the largest elecrical company in Britain, with a market capitalisation of £14·5 million, with AEI second (£9·8 million) and English Electric third (approximately £5 million). Together, these companies accounted for at least sixty per cent of the market for power station equipment, as well as manufacturing the complete range of electrical products, from domestic appliances right through to turbo-generators. They were the only British manufacturers capable of performing this broad role in the electrical market because after the turn of the century most companies had concentrated on various specialised products. This had been a response to the force of competition from the new generation, but by the inter-war period it allowed a large number of companies to survive on the basis of a relatively narrow product range. It was a trend which undermined the level of profitability within the industry generally, encouraging severe price competition even when demand was buoyant, and, as BEAMA (1927) warned, such a market structure could only prove damaging to the long-term prospects for British electrical manufacturing. As we saw earlier, the mid-1920s had also seen the dissolution of the price rings created by BEAMA in 1911, and it was only after 1931 that manufacturers agreed once again to their reintroduction.

Although still only a medium-sized electrical company, with an issued capital of £500,000 in 1930, it would be unrealistic to write off Ferranti as a minnow competing in the same markets as the multi-million pound businesses like GEC, English Electric and AEI. Following the capital reconstruction of 1923, Ferranti had been forced into another restructuring in 1929 to provide for the increase in business. The 1923 scheme, of course, had not raised any fresh capital, because it had been designed simply to restore Dr Ferranti and Vincent as the majority shareholders, but the appendix reveals how, despite the depression of the early 1920s, Ferranti had improved its profitability beyond all recognition when compared to the pre-war years. This surplus cash was used in a variety of ways: firstly, by 1928 all the debenture stock created prior to 1914 had been redeemed, reducing the burden of interest payments; secondly, the General Reserve was built up to £270,000 by June 1929, providing the resources for future expansion and, thirdly, between January 1923 and June 1925 the company purchased a holding of £76,000 in five per cent War Loan, selling it off at the end of 1926 to finance the Stalybridge Radio Department. This prudent and imaginative policy provided for the expansion and diversification of the 1920s, but by 1929 the bank overdraft had returned to haunt the Board after having been kept within manageable limits since 1920. After the war there had been a series of consolidations in the banking system, and Parr's had been absorbed into the London County & Westminster Bank in 1918. The name had been changed to the Westminster Bank by 1923, but Ferranti's relationship with the new organisation re-mained as close as it had been in the late 1890s, and overdraft facilities were provided frequently in the 1920s. This was extremely useful in a period of expansion, but by 1929 the overdraft stood at £97,000 and the Board decided to raise fresh capital to reduce this burden. The Manchester legal firm of Skelton & Company were commissioned to prepare a scheme 'to provide for future require-ments' and it was decided to raise the nominal capital from £300,000 to £900,000. This was achieved by the creation of 300,000 new seven per cent Cumulative Preference Shares and 600,000 Ordinary Shares, and the conversion of all shares from a value of £1 to 10s. As in 1923, the Ferranti family was given all the Ordinary Shares issued (400,000), through the capitalisation of £200,000 from the General Reserve Account, but the stockbrokers Foster & Braithwaite were able to sell publicly all the 250,000 Preference Shares issued. Voting rights were once again attached to all the Preference Shares, but after the 1929 reconstruction Dr Ferranti and Vincent then held forty-one per cent of the equity. By providing an increase in the issued capital, from £206,166 to £446,166, this also ensured that the much-needed cash for the CEB contracts and the new Valve Department would be available, at the same time maintaining Ferranti family control over a large proportion of the voting stock.

The timing of Ferranti's reconstruction, in June 1929, seems to have been rather dangerous because the major boom in shares had already started to subside, but still Foster & Braithwaite had placed all the 250,000 Preference Shares. This is a firm indicator of the company's turn-round from the events of 1901, when only half of the 120,000 Preference Shares issued had been sold, and the collapse of 1903. Of course, the added attraction of the voting right would have influenced the sale, but investors were clearly impressed with both the improvements in profits since the war and the potential for growth in an organisation with large order-books and plans for diversification into the radio set business. One must emphasise that, in fact, Ferranti's gross return on capital employed had actually fallen in the 1920s, from around twenty per cent in 1920 to ten per cent in 1930, but this was still a better performance than the company's main rivals. During the 1920s electrical companies had suffered badly from the ruinous price competition, and BEAMA (1927) records that between 1925 and 1928 the proportion of net profits to total capital for the eleven largest manufacturers in the industry (including Ferranti) had fallen from 9·35 per cent to 7·7 per cent. In contrast, a similar figure for Ferranti would record a fall from 15·4 per cent to nine per cent, while even in 1930, when the world depression was beginning to bite and the company's capital employed had been increased by over £260,000 on the 1928 figure, they could claim a net return on total capital of 7·9 per cent. These figures illustrate how, with output rising to £1·2 million, Ferranti had made a real recovery from past difficulties. By allowing Dr Ferranti and his engineering team the resources to improve the competitive standing of the company's products, the Board had ensured that the opportunities of the era would be exploited to the fullest extent within the confines of their financial resources. Tait and Whittaker had played their roles in this progress, and marketing was maintained as an important managerial tool, but the key to Ferranti's performance must remain the technology-led strategy formulated by its founder.

One could not discuss the company's financial results in any period of Dr Ferranti's life without emphasising the contribution of the Meter Department. Although the Transformer and Radio Departments grabbed all the headlines in the 1920s, the Meter Department continued to provide the profits to subsidise these risky ventures, accounting for at least half of company output in each year between 1918 and 1930. Dr Ferranti and the Chief Meter Engineer, G. Wall, were responsible for a series of improvements in the design of both AC and DC meters, the former taking out twenty-eight patents covering this line alone in the last twelve years of his life. The growing use of AC in both home and export markets also led to greater attention being paid to this particular type, and the introduction of the FD AC meter in 1919 contributed significantly to the progress made in expanding departmental sales from £91,000 in 1918 to £575,000 in 1930. Ferranti had maintained the meter price ring in effective order in the 1920s, despite its role in breaking up the transformer agreements, and this allowed the company to make an annual trading profit on this business of between twenty-two and thirty-four per cent per annum over these years. This gross figure always exceeded the profit figure recorded in the balance sheets and, as in previous years, without the Meter Department Ferranti could never have afforded its expansionist policies. An additional point relates to the company's position *vis-à-vis* its much larger rivals like the big three in that, while GEC, AEI and English Electric could claim to be stronger in the electrical market overall, in its chosen fields Ferranti could compete with any rival. Of course, as we have already noted, in the radio-set market the competition of specialist manufacturers prevented greater penetration, but in meters and transformers the company accounted for large shares of the business in the home market. It is always difficult to make exact comparisons between company output statistics and the national census returns, but using these figures in 1930 Ferranti could claim thirty-four per cent of British house-service meter production (of £1·7 million) and fourteen per cent of the static transformer business (of £2·9 million). As we saw earlier, with regard to the larger transformers for the CEB, Ferranti had claimed over one-third of this market, substantiating the general point that the company had nothing to fear from any British manufacturer in the markets on which it depended.

Ferranti by 1930 was a sound business, compared both to its rivals and to its performance prior to 1914. We have seen in this chapter how it had initially streamlined its electrical business while completing the munitions contracts and, using the opportunity created by the growth of the electricity supply network and the elimination of foreign competition from the home market, Dr Ferranti had then exploited his generally-acknowledged expertise in transformer design to establish his company's reputation as the major British manufacturer of the largest units. He had also ensured that the Meter Department would continue to provide the funds for future expansion and in the mid-1920s Ferranti was led into one of the most exciting new industries, radio. This was to result in a radical switch in technological orientation, with risky ventures into valve manufacture and more fundamental research and development work on other electronics projects. Vincent maintained this switch into new technologies, employing the technology-led strategy his father had built into the structure of the company to good effect over the next few decades in building Ferranti into a diversified electrical engineering and electronics manufacturer (Wilson, 1984). Dr Ferranti's contribution to these developments has been well charted and the Board had recognised this in the 1923 reconstruction of the capital, but finally in 1928 he was given pride of place in the company after Tait's retirement. The Chairman had actually been involved in a financial scandal associated with some dubious investments in an East African soda mine, as Davenport-Hines (1986) reveals, and by 1925 this had all but broken his reputation. He remained at the head of Ferranti for a further three years but, after suffering a series of setbacks to his health, he finally retired in May 1928. Tait had contributed enormously to the company's recovery since 1903 and he left it in a much more secure position than he had found it, but by 1928 he could no longer cope with the rigours of business. In his place, Dr Ferranti was unanimously elected to take over as Chairman, having already become both majority shareholder in 1923 and the company's leading source of new ideas since 1914. This would have been no idle gesture by men like Whittaker and Cooper, but it was clear that Ferranti was not to return to the type of family company it had been up to 1903. By the 1920s Dr Ferranti had been persuaded by harsh experience that balance was all-important in business and, as we have already noted, this was to be a key feature of Ferranti's progress. When he died, rather suddenly in January 1930, Dr Ferranti left a company well-suited to the demands of a technologically-progressive industry, a lasting achievement which stood the test of time.

6

The elder statesman

When Dr Ferranti died on 13 January 1930, at the age of sixty-five, the British electrical industry lost one of its most respected figures. Having been a manufacturer virtually from the birth of the industry in 1882 right up to the implementation of the 1926 Electricity (Supply) Act, he had worked consistently and effectively in pursuit of the 'All Electric' ideal. As his Royal Society obituary stated, 'in his own view his life's mission was to spread the gospel of the use of electricity for every possible purpose ... It is clear that he exerted great influence on contemporary thought in electrical engineering all over the world.' His passionate belief in this 'mission' certainly inspired him to great efforts, and throughout a forty-nine year career he committed his considerable foresight and energy to the improvement of electrical technology. While one might criticise the economic rationale behind some of his work, there always ran a consistent thread of thought through the large number of projects in which he became involved — as the *Electrical Review* of November 1910 claimed, 'how to shape the project in hand to the economic needs of the future'. He was, above all else, a man with vision, a theme in his work which can be discerned just as much in the development of the smallest meter as in the ambitious Deptford project. His real problem lay in matching the ideas to an unresponsive environment for electrical technology in Britain, and two highly-publicised commercial failures (at Deptford and with his company in 1903) feature strongly in a life spent struggling against the obstacles placed in the path of greater electrification. Dr Ferranti could never be described as a commercial engineer, and it is no coincidence that only when chartered accountants were brought onto the Board in 1905 did his company begin to exploit successfully the profit potential in the products he designed. At the same time, his technology-led strategy provided the basis for an expanding business which has contributed extensively to the development of British electrical and electronic engineering for over one hundred years, and clearly this is how Dr Ferranti should best be remembered. Just as with business management, labour relations never proved to be his *forte* but, in establishing a company which increased its number of employees considerably in the inter-war years, he left a valuable legacy to a depressed area of Lancashire. This, indeed, might be regarded as his greatest achievement. Dr Ferranti was one of the most eminent British electrical pioneers. He worked hard for such organisations as the British Electrical Development Association in the 1920s and many honours were bestowed upon him in recognition of a lifetime dedicated to the 'All Electric' ideal. Yet, in Ferranti Limited a lasting monument to his work as an engineer had been built, reminding us of the enterprise shown during the electrical industry's difficult early decades.

When Dr Ferranti first experimented with magnets in 1878, he was immediately convinced that electricity ought to be the energy source of the future and, over the following fifty-two years, more work was put into fulfilling this aim than any other of his multifarious ideas. Central to his hopes for an 'All Electric' future was the need for more efficient and cheaper generating and transmitting machinery and, as far as electrical engineering was concerned, Dr Ferranti's ideas for new types of generators, steam engines, turbines, transformers, switchgear and cables were of great significance in their own time. We have noted at several points in this story how his designs laid the basis for future work: the Grosvenor Gallery and Deptford alternators were the first large-scale generators of high-tension AC; the resuperheating turbine provided the increased power required by the 1920s; Ferranti produced the most advanced transformers in Britain for the national grid; the switchgear of the mid-1890s established the high levels of safety later insisted upon by British power stations, and the 10,000 V cables for LESCo were the first effective means of carrying such high pressures. These were just some of the most outstanding of his contributions to British, and general, power-station technology, on top of which the construction of the first modern power station at Deptford and the description of a grid network in his 1910 presidential address to the IEE added further to his reputation as a pioneer of what was eventually to become standard practice in the electricity supply industry. Perhaps he never achieved the practical successes accredited to Charles Merz, but what Hannah (1979) describes as Dr Ferranti's 'futuristic vision of engineering genius' can be regarded as an apt description of his work. The grid network was very much the fulfilment of the 'Ferranti Dream', a phrase coined as early as 1890 by the *Electrical Review* to describe his hopes of rationalising the electricity supply industry by concentrating production in high-tension AC super-power stations built where land was less expensive, water available, and fuel cheaper to transport. Again, he was not the only advocate of this essential reform, but it was no coincidence that these were the principles eventually implemented in the 1920s and the 1930s, when politicians finally accepted the logic behind them. It was highly appropriate that his 10,000 V LESCo cables should only be replaced in 1932, to make way for the CEB's 132,000 V mains, an event marked by an article in the *Electrical Times* of December 1932 headed 'Sic Transit ...'.

It is clear that Dr Ferranti contributed effectively to the development of electrical engineering during his lifetime but, as we have noted in earlier chapters, one of his least spectacular, yet most enduring, influences came in meter design. While historians have concentrated on his larger projects, like Deptford and the steam alternator, both of which resulted in heavy financial losses, one should always remember that Dr Ferranti developed both an accurate and reliable meter and a highly profitable business in these instruments. From his mercury-motor meter of 1885 through to the FD meter of the 1920s, Ferranti meters consistently accounted for large shares of both home sales and British exports. The business may not have been as huge as that in electric bulbs or generators, but for Ferranti it provided the profits for the ventures into high-risk areas like radio and subsidised, for example, the transformer development programme of the 1920s. This was possible for four main reasons: Dr Ferranti ensured that the design remained competitive; he introduced mass-production techniques to keep costs down; after 1905 an extensive agency network was created to increase exports, and by 1907 a price ring had been formed. The Meter Department was one of Dr Ferranti's most outstanding technical and commercial achievements. In the standardisation of production techniques, in particular, the venture was regarded as one of the best contemporary examples of modern works design, with the extensive use of American light machine tools and a carefully-planned flow-line organisation. This attention to detail, in both the design and manufacture of meters, paid handsome dividends for the company, and without the financial resources accumulated from this department it is clear that Ferranti could never have afforded its expansion and diversification strategies. In this context, the meter is given a place of honour in the company's history and no other product played such a vital role for so long.

While the Meter Department provided the cash for the company's more ambitious strategies, it is essential to explain why this product remained competitive and why Ferranti ventured into new products. Here again, we must emphasise the importance of Dr Ferranti's techology-led strategy; how, from his earliest days as a manufacturer through to his last years as Ferranti's Technical Manager, he continually strived for better designs. Dr Ferranti was an incorrigible developer and this is why his relationships with businessmen like Francis Ince, J. S. Forbes and A. W. Tait were always soured by their desire to control such impulses. In an industry with a high rate of technological obsolescence, product development was an essential aspect of any electrical manufacturer's approach to the market, and by the inter-war years Dr Ferranti had built team research into the structure of his company. We have seen in the last chapter how a post of Technical Manager was specifically created for him in 1918 and, using this authority, he established research departments for each of the main product lines. The policy of hiring academic scientists to act as consultants to the Radio Department complemented this attempt to ensure that a high level of technical input went into all Ferranti products. Many leading manufacturers in the electrical, and other, industries were, of course, pursuing similar policies by this time, as Sanderson (1972) notes, but Dr Ferranti's technology-led strategy provided the basis for a successful business after the First World War as the market expanded to create new opportunities. It also induced engineers of some ability to join Ferranti, and during his lifetime Dr Ferranti was able to pass on the benefits of his inspiration to men who later contributed extensively to the industry. Men like H. W. Clothier, S. Ferguson and S. Pailin in switchgear, E. T. Norris in transformers, and Charles Sparks in electricity supply all worked intimately with Dr Ferranti, and his approach to team-work brought out the talents they possessed. What better way therefore for the IEE to commemorate Dr Ferranti's contribution to electrical engineering than to establish after his death a Ferranti Research

Scholarship for promising young engineers, a scheme he would have heartily encouraged.

Although team-work became a key feature of Ferranti's technology-led strategy, Dr Ferranti would have been disgusted if anybody had ever interpreted this in terms of industrial democracy. While he was always working on the shop-floor and he liked to keep on friendly terms with the workforce, Dr Ferranti retained a dislike of trade-union interference in management throughout his life. We saw in Chapter 3 how his uncompromising stand in the 1897 engineers' lock-out created an unfavourable impression in the local area and, after his return to the Crown Works in 1914, a similar aversion to trade-unions still crept through. When the company decided to create a Works Committee in 1917, to improve communication between management and the expanding workforce – details of which can be found in the Appendix – he complained in a letter to Vincent that it only wasted valuable working hours. At the end of the war he also commented: 'I look upon the labour position as most serious as if we win the present war we have a worse enemy to fight in the shape of a misled lower class who want to nationalise everything and tax people who have or are capable of making anything out of existence.' This reflects some of the general concern voiced by contemporaries over the alleged militancy of the trade-union movement after 1918, a militancy which was fuelled by the rampant post-war inflation and the government's reluctance to implement any of the welfare schemes promised after the end of hostilities. In the engineering industry, just as in the 1890s, a further issue was the right to manage the shop-floor, culminating in the second great dispute between the Engineering Employers' Federation (EEF) and the engineers' union (now renamed the Amalgamated Engineering Union) in 1922 (Wigham, 1972). Dr Ferranti never played as prominent a role in this fight as he had done in 1897, but his firm stood solidly behind the EEF's forceful campaign and the AEU was defeated. The company was also involved in a series of smaller disputes in 1918 and 1919 concerning its reluctance to return to the *status quo* in the use of skilled labour. The trade-unions had agreed with the government to extensive dilution in 1915, on condition that employers would restore traditional workshop practices after the war, but by February 1919 the Crown Works had been brought to a standstill by Ferranti's refusal to replace unskilled female labour with skilled engineers in the production of transformers. It was only the intervention of the Joint Committee of Engineering and Kindred

Trades, part of the industrial relations machinery set up after the Whitley Report, which forced Ferranti to abide by the 1915 agreements, but Dr Ferranti regarded this as unwarrantable intervention in the company's affairs.

One of the methods employed by engineering employers of this period to reduce the influence of trade-unions was the introduction of company welfare policies, and after 1915 there does appear to have been a more professional approach adopted in the Crown Works towards labour relations. Dr Ferranti and the Board were well aware that if the company could provide certain services, as an alternative to those of trade-unions, more effective control of the workforce might be possible. This is reflected in the creation of a Welfare Department in 1917, for the discussion of industrial grievances, the establishment of the Ferranti Recreation Club in the same year, and the purchase of over thirty cottages within the vicinity of the works to house foremen and senior employees. When the large-scale employment of female labour started in 1915, Dr Ferranti recruited a close friend of the family, Olivia Forbes, to become the company's first Lady Supervisor. She remained with Ferranti for almost forty years and, besides her work in recruiting and supervising the women employed by the company, she also lectured to the Institute of Industrial Welfare Workers on the welfare policies pursued at the Crown Works. The Appendix reveals that the workforce expanded by almost 140 per cent between 1919 and 1930, from 2,014 to 4,773, and this prompted further structural changes on the labour relations side with the creation of a new post in 1925, Works Manager. The first Works Manager was J. W. Davies, a man who forged a close link between the company and trade-union leaders in the Manchester area. He compensated for some of the less diplomatic statements made by Dr Ferranti on the issue of labour relations, indicating the extensive professionalisation of management within the company by the 1920s. As we have noted in previous chapters, while Dr Ferranti was appointed Technical Manager, his fellow-directors were responsible for the company's finances, salesmen were introduced into the organisation, and on the labour management side a Lady Supervisor and a Works Manager controlled relations with the workforce. This functional specialisation was symptomatic of the growing scale of operations within the business, especially after the expansion into a second factory in 1926. On the labour relations side, Ferranti appeared to have experienced few industrial disputes after 1922, but the spectre of unemployment which haunted the Hollinwood area in the inter-war years may well have contributed to this situation. Ferranti were regarded as no better and no worse than the average medium-sized engineering company in this period, and even though the Ferranti family were ever-present within the works the old-style paternalism of the industrial North-west never seems to have been a popular policy within the company. The Board preferred to conduct its labour relations through professional managers and, in view of Dr Ferranti's hostility to trade-unions, this worked well enough in controlling an expanding workforce.

Ferranti's growing sophistication as a business venture is an interesting illustration of the general trends between the wars with regard to the use of new managerial techniques. At the same time, leadership is the prime quality which usually dictates whether or not a business will survive, and Dr Ferranti's ethos had been firmly embedded into the structure of the company to ensure that provision for future growth had been made. His son Vincent extended the technology-led strategy by leading Ferranti into some of the vital growth sectors of the twentieth century, but one must always remember that the foundations for this strategy were laid during Dr Ferranti's lifetime. Ferranti is still one of the leading companies in the electrical industry today, and Dr Ferranti's grandson, Basil, remains Chairman (albeit with non-executive status) of a group which employs over 20,000 people around the world. Although it has pulled out of the declining meter and transformer businesses in the last decade, Ferranti has a world-wide reputation in such fields as avionics, computer systems and microcircuits. This product range has been built up by investing a high proportion of its profits in development programmes and technological innovation, putting the company into Britain's technical vanguard. Dr Ferranti left behind him a company based on a balanced approach between the desire for technical excellence and the need to be financially solvent, a philosophy which had evolved slowly over his lifetime as a manufac-turer. By the 1920s, Dr Ferranti had clearly been obliged to modify his fierce individualism of the years up to 1903 and, in allowing professional businessmen the opportunity to control the key managerial tools of finance and sales, he provided Ferranti with the balanced outlook which has been a key to its success over much of the last two generations.

Ferranti Limited, as we have already noted, was Dr Ferranti's greatest achievement, but in the 1920s he received much more credit for his general work in the industry. After having been accorded the privilege of serving the first double term as President of the IEE in 1910 and 1911, in 1924 he was awarded its highest honour, the Faraday Medal, only the third man ever to receive this distinction. The ultimate accolade from the scientific fraternity came in 1927 when he was elected a Fellow of the Royal Society, a position regarded as a worthy acknowledgement of his engineering prowess. Even in the USA he was held in similarly high esteem, having been made an honorary member of the American IEE as early as 1911, one of the only five such persons in the world at that time. This high regard for a man with an infectious enthusiasm for the 'All Electric' ideal reflected his position as an elder statesman in the industry. It was this eminence which prompted his election as President of the British Electrical Development Association (BEDA) in 1926 to act as the leader of publicity campaigns to extend the use of electricity in the home. He also assisted Caroline Haslett in her work as creator of the Electrical Association for Women (EAW), another useful vehicle for popularising the advantages of electrical gadgets. Unlike many of his counterparts in industry in the 1920s, Dr Ferranti never used his reputation to enter politics. He preferred instead to concentrate on engineering and the struggle to create an 'All Electric' age. As he stated in 1924, after receiving the Faraday Medal: 'We cannot have too much electricity. We want it everywhere, and we have not anything like enough of it.' This was the philosophy which underpinned his work outside the Crown Works, and by the 1920s many more people, from housewives to politicians, were convinced that this message should be implemented as soon as possible.

It is difficult to establish the full effect of such publicity campaigns on the British public. Certainly, as Hannah (1979) explains, BEDA lacked the funds to make a bigger impact on the market, while the EAW concentrated its attentions on the middle classes, rather than the mass market. Electricity consumption continued to rise, from 3,569 GWh in 1918 to 9,169 GWh in 1930, and by 1938 sixty-six per cent of all houses in Britain were wired for electricity, compared to only six per cent in 1918. Nevertheless, even in the inter-war years the electrification of the house and industry proved to be a slow process. Dr Ferranti put his heart and soul into the publicity campaigns and, quite typically, he tried to lead by example in making

Baslow Hall one of the earliest 'All Electric' homes. Gertrude also assisted in this work, creating the North Wales branch of the EAW in 1926 and contributing an article to the *Electrical Review* in 1928 on the advantages of domestic electrical appliances. This prominence in the public mind even provided Ferranti with an opening in the domestic appliance market, after Dr Ferranti designed in 1927 what became one of the best-known electric fires in Britain. Unfortunately, up to Dr Ferranti's death, annual sales in Ferranti consumer products never exceeded £22,000 and only the introduction of protective duties in 1931 could exclude the fierce competition from American producers. The story of this venture belongs more to Vincent's ear as Chairman, when sales did increase and the product range was extended to include clocks and water-heaters, but up to 1930 it typifies the continued difficulties in expanding the electricity market. The grid's start-up, of course, also provided a huge boost to the industry, but it is clear that greater effort would have to be made before the 'All Electric' age became a reality, and right up to his last years Dr Ferranti was pleading for more money to be spent on publicity within the industry as a whole.

The life of Dr Ferranti had been spent in pursuit of the 'All Electric' ideal. Tragically, after suffering from chronic rheumatism and severe problems with his prostate gland throughout the 1920s, he was to die before the CEB grid went into operation. He was actually taken ill while on holiday in Switzerland, where the cold air aggravated his prostate gland to such an extent that he was rushed to a hospital in Chur, where it was removed. This operation proved only partially successful, and the day after a clot of blood found its way to his heart, killing him instantly. Even when being wheeled on a stretcher along the railway platform at Chur, though, he insisted on stopping for some time by the side of an electric locomotive to explain its workings to his youngest son, Denis. Not even severe pain could quench his electrical interests, just as he had refused to allow any obstacle to thwart his efforts during his career. This indomitable spirit inspired him, and those around him, to a fuller realisation of the prospects for electricity, and in the process British power-station technology made rapid strides, while an important electrical company was founded which prospered on the basis of Dr Ferranti's love of engineering. British culture has bred an aversion to praising 'Captains of Industry', while in other countries they are lauded for their achievements. One must be wary of eulogising their role in the economy, and in Dr Ferranti's case we have been careful to describe the failures as well as the successes, but in creating Ferranti Limited and by improving power-station technology he played a significant role in the evolution of the British electrical industry. It is for these reasons that he deserves credit for being one of Britain's most innovative engineering industrialists.

Appendix

This financial information on the performance of Ferranti's businesses is provided to give the reader a clear view of certain important indicators: output, capital employed, net profits and labour force. I have chosen to use net profits because the earlier balance sheets are very poor indicators of gross returns. It also gives a clear indication of the amount remaining for reinvestment. (The figures in parenthesis are losses.) There are several gaps in the figures (denoted by the use of NA for not available) but this is inevitable given the company's private status.

	Output £	Capital employed £	Net profits £	Labour (as at 3 January)
1889	54,300	NA	NA	NA
1890	86,900	NA	5,673	NA
1891	31,800	NA	(5,071)	NA
1892	35,222	NA	(8,476)	NA
1893	39,900	NA	1,218	NA
1894	50,500	106,717	1,118	NA
1895	63,220	128,437	3,964	NA
1896	99,245	133,201	3,561	NA
1897	NA	157,814	771	NA
1898	NA	192,413	7,915	NA
1899	NA	200,948	17,635	NA
1900	NA	224,371	15,275	NA
1901	NA	263,109	25,160	NA
1902	NA	328,755	3,875	NA
1903	NA	NA	(58,557)	NA
1904	NA	NA	(40,114)	NA

	Output £	Capital employed £	Net profits £	Labour (as at 3 January)
1905*	NA	NA	(7,334)	556
1906	NA	246,320	(169)	653
1907	128,212	250,115	2,092	762
1908	129,465	265,155	3,275	656
1909	136,185	263,853	307	797
1910	160,162	262,824	2,991	869
1911	227,189	263,519	4,166	1,168
1912	261,906	278,937	7,129	1,584
1913	294,639	305,337	10,062	1,752
1914	267,994	295,918	1,598	1,432
1915	NA	296,721	(1,022)	1,216
1916	NA	299,395	3,546	1,880
1917	NA	254,928	7,650	2,555
1918	207,704	266,002	35,174	2,994
1919	461,795	279,362	18,251	2,014
1920	799,955	339,824	36,269	1,872
1921	666,818	403,489	36,269	2,190
1922	489,146	384,087	24,371	1,783
1923	566,675	392,969	36,362	2,056
1924	744,654	416,526	55,546	2,340
1925	834,704	438,212	67,644	2,492
1926	916,004	485,120	59,766	3,321
1927	989,582	561,455	35,982	3,728
1928	NA	606,610	54,616	4,009
1929	1,113,527*	741,321	66,976	4,200
1930	1,206,417	868,921	68,749	4,773
1931	1,186,786	856,678	60,689	4,322
1932	1,159,837	930,750	54,905	4,663
1933	1,191,471	858,308	57,869	3,672
1934	1,101,490	853,654	68,245	3,823
1935	1,372,012	1,058,451	55,902	5,394
1936	1,456,844	1,166,039	57,438	5,160
1937	1,511,263	1,374,543	64,923	6,170
1938	1,889,980	1,372,783	68,263	NA
1939	1,639,872	1,212,603	79,383	5,703

* Year to June hereafter.

Note If one was to conduct a similar exercise for Ferranti plc, the relevant figures would be (for 1985): turnover of £568 million and operating profit of £50·2 million.

Sources and further reading

A note on sources

Readers are referred to my doctoral thesis, *The Ferrantis and the Growth of the Electrical Industry, 1882–1952* (Manchester, 1980) for a full list of the sources available in the Ferranti Archives. This depository provided much of the material on the Ferranti family and its business activities, but a comprehensive search of such journals as *Electrician* (1872–1930), the *Electrical Review* (1878–1930) and *Wireless World* (1924–30) has also been conducted to place the story intơ a general context. Parliamentary Papers have been useful in assessing the performance of the electrical industry, including *The Departmental Committee on the Engineering Trades after the War* (Cd. 9073, 1918), the *Report on the Engineering Trades (New Industries) Committee* (Cd. 9226, 1918) and the Monopolies and Restrictive Practices Commission *Report on the Supply and Export of Electrical and Allied Machinery and Plant* (1956). One of the best primary sources on this subject might be Ferranti's two main speeches, in 1910 and 1928: 'Coal conservation, home-grown food, and the better utilisation of our labour', *Journal of the IEE* Vol. 46 (1911), and 'Electricity in the service of man', *Journal of the IEE*, Vol. 67 (1929).

Further reading

There is a growing literature on the electrical industry's evolution, and a list of the major authorities used in this book is given below.

Appleyard, R. *The History of the IEE, 1871–1931*, IEE, 1939.

Bailey, F. *The Life and Work of S. Z. de Ferranti*, Electrical Power Engineers Association, 1932.

Ballin, H. *The Organisation of Electricity Supply in Great Britain*, Electrical Press, 1946.

Byatt, I. C. R. *The British Electrical Industry, 1875–1914*, Clarendon Press, 1979.

Catterall, R. E. 'Electrical engineering', in N. K. Buxton and D. H. Aldcroft (eds.), *British Industry between the Wars*, Scolar Press, 1979.

Clark, R. W. *Edison: the man who made the future*, London, 1977.

Clothier, H. W. *Switchgear Stages*, Reyrolle, 1933.

Cottrell, P. *Industrial Finance, 1830–1914*, Methuen, 1980.

Davonport-Hines, R.P.T. *Dudley Docker. The Life and Times of a Trade Warrior*, Cambridge University Press, 1984.
— 'A.W. Tait', *Dictionary of Business Biography*, Vol.5, 1986.
— 'Two autobiographical fragments by Hugo Hirst', *Business History*, Vol.28, 1986.
G.Z. de Ferranti and R. Ince, *The Life and Letters of S.Z. de Ferranti, Williams & Norgate, 1932*.
Fleming, J.A. *Fifty Years of Electricity*, London, 1930.
Hannah, L. *Electricity before Nationalisation*, Macmillan, 1979.
Hughes, T.P. 'The British electrical industry lag, 1882–1888', *Technology and Culture*, 1962.
— 'The bold conception of S.Z. de Ferranti', *Consulting Engineer*, 1963.
— *Sperry. Inventor and Engineer*, Johns Hopkins University Press, 1971.
— *Networks of Power*, Baltimore Press, 1983.
Irving, R.J. 'New industries for old? Some investment decisions of Sir W.G. Armstrong, Whitworth & Co. Ltd., 1900–1914', *Business History*, Vol.17, 1975.
Jones, R. and Marriott, O. *Anatomy of a Merger*, Pan, 1972.
Johnson, J.H. and Randall, W.L. *Colonel Crompton*, Longmans Green, 1945.
Kennedy, W.P. 'Foreign investment, trade and growth in the UK, 1870–1913', *Explorations in Economic History*, Vol.11, 1974.
Parsons, R. *The Early Days of the Power Station Industry*, Cambridge University Press, 1939.
Passer, H.C. *The Electrical Manufacturers*, Cambridge, Mass., 1954.
Payne, P.L. 'The emergence of the large scale company in Britain, 1870–1914', *Economic History Review*, 2nd Series, xx, 1967.
Ponting, K.G. 'The textile experiments of S.Z. de Ferranti', *Textile History*, Vol.4, 1973.
Sanderson, M. *Universities and British Industry, 1850–1970*, Routledge & Kegan Paul, 1972.
Saul, S.B. 'The American impact on British industry, 1870–1914', *Business History*, Vol.3, 1960.
— (ed.), *Technological Change: United States and Britain in the Nineteenth Century*, Methuen, 1970.
Sturmey, S.G. *The Economic Development of Radio*, Duckworth, 1958.
Whyte, A.G. *Forty Years of Electrical Progress. The Story of the GEC*, Ernest Benn, 1930.
Wigham, E. *The Power to Manage*, MacMillan, 1972.
Wiener, M. *English Culture and the Decline of the Industrial Spirit, 1850–1980*, Cambridge University Press, 1981.
Wilson, J.F. 'A strategy of expansion and combination: Dick Kerr & Co., 1897–1914', *Business History*, Vol.27, 1985.
— 'Sir Gerard Vincent Sebastian de Ferranti', *Dictionary of Business Biography*, Vol.2, 1984.

Index